Anonymous

The Castle Line Atlas of South Africa

A Series of 16 Plates, Printed in Colour, Containing 30 Maps and Diagrams

Anonymous

The Castle Line Atlas of South Africa
A Series of 16 Plates, Printed in Colour, Containing 30 Maps and Diagrams

ISBN/EAN: 9783337117498

Printed in Europe, USA, Canada, Australia, Japan

Cover: Foto ©berggeist007 / pixelio.de

More available books at **www.hansebooks.com**

OF

SOUTH AFRICA,

A SERIES OF 16 PLATES, PRINTED IN COLOUR,
CONTAINING 30 MAPS AND DIAGRAMS.

WITH AN ACCOUNT OF THE GEOGRAPHICAL FEATURES,
THE CLIMATE, THE MINERAL AND OTHER RESOURCES,
AND THE HISTORY OF SOUTH AFRICA.

AND AN INDEX OF OVER 6,000 NAMES.

LONDON
DONALD CURRIE & CO.,
1, 2, 3, & 4 FENCHURCH STREET, E.C.
1895.

CONTENTS.

"SUNNY SOUTH AFRICA."

PAGE

CHAP.

SOUTH AFRICAN RAILWAYS (see Maps 3, 4, 7, 8, 9, 12 and 13).   4

I. A LAND OF DIAMONDS AND GOLD (see Maps 4, 5*b*, 13, 14 and 15)   5

II. THE COUNTRY OF THE VELDT AND THE KARROO (see Maps 4, 5*a*, 10 (inset), 11, 12 (inset), and all the general maps)   8

III. A LAND OF SUNSHINE AND HEALTH (see Maps 1, 2, 3 and 6*a*)   16

IV. THE SPORTSMAN'S PARADISE (see Map 4)   21

V. THE PEOPLE OF SOUTH AFRICA (see Map 6*b*)   25

VI. THE MAKERS OF SOUTH AFRICA (see Maps 2, 3 and 4)   29

VII. THE STORY OF SOUTH AFRICA (see Maps 7, 12, 15 and 16)   33

LIST OF MAPS.

1. THE BRITISH EMPIRE, ON MERCATOR'S PROJECTION.

2. AFRICA, WITH INSETS OF MADEIRA AND THE CANARY ISLANDS, SHEWING CASTLE LINE MAIL AND INTERMEDIATE ROUTES.

3. CENTRAL AND SOUTH AFRICA, SHEWING COMMUNICATIONS AND MISSION STATIONS.

4. AFRICA SOUTH OF THE ZAMBESI, POLITICAL AND INDUSTRIAL.

5. SOUTH AFRICA, PHYSICAL; SOUTH AFRICA, GEOLOGICAL.

6. SOUTH AFRICA, RAINFALL; SOUTH AFRICA, ETHNOLOGICAL.

7. CAPE COLONY IN DIVISIONS, WITH BASUTOLAND AND THE ORANGE FREE STATE.

8. SOUTH AFRICA—WESTERN SHEET, SHEWING WESTERN CAPE COLONY.

9. SOUTH AFRICA—CENTRAL SHEET, SHEWING CENTRAL CAPE COLONY, AND PART OF THE ORANGE FREE STATE.

10. SOUTH AFRICA — EASTERN SHEET, SHEWING EASTERN CAPE COLONY, NATAL, BASUTOLAND, ZULULAND, AND PART OF THE ORANGE FREE STATE, WITH A PLAN OF PIETERMARITZBURG.

11. THE CAPE PENINSULA, AND PLANS OF CAPETOWN, PORT ELIZABETH, AND DURBAN.

12. SOUTH AFRICAN REPUBLIC AND THE ORANGE FREE STATE, WITH PLAN OF PRETORIA.

13. SOUTH AFRICA—NORTH SHEET, SHEWING THE TRANSVAAL GOLD FIELDS, WITH A PLAN OF JOHANNESBURG AND ITS SUBURBS.

14. PLANS OF THE DE KAAP AND MOODIE, KOMATI, WITWATERSRAND, AND KLERKSDORP GOLD FIELDS.

15. MATABELELAND AND MASHONALAND.

16. EAST CENTRAL AFRICA, INCLUDING BRITISH CENTRAL AFRICA AND NYASSALAND.

SOUTH AFRICAN RAILWAYS.

CAPE COLONY.—The three main or trunk systems of railways in the Cape Colony are called the Western, Midland, and Eastern.

The Western Railway, with its starting point in Capetown, and the Midland Railway from Port Elizabeth, are worked as a single trunk system, 839 miles in length, the connection between the two systems being at De Aar, 501 miles from Capetown, and 339 from Port Elizabeth. Through trains are also despatched from both Capetown and Port Elizabeth to Kimberley, the centre of the Diamond Fields, 647 miles from the former, and 485 miles from the latter port; and the line has been extended to Vryburg and to Mafeking. Another line, with junctions at De Aar and Naauwpoort, has been completed, via Bloemfontein to Johannesburg and Pretoria. On the Western portion of the line the principal stations are Durban Road, Paarl, Wellington, Ceres Road, Worcester, Tonw's River, Beaufort West, and De Aar. There are branch lines to Malmesbury, and to Wynberg, Kalk Bay, and Simon's Town; and another, 42 miles in length, from Worcester to Robertson and Roodewall (Kogmans Kloof)—the station for Montague—has been opened for traffic.

The chief stations on the Midland Railway from Port Elizabeth are Alicedale Junction, Cookhouse, Cradock, Middleburg Road, and Naauwpoort Junction, where it joins an extension to Bloemfontein, Kronstad, and Viljoen's Drift (Vaal River), Johannesburg, and Pretoria. Branch lines run to Uitenhage, and Graaf-Reinet, Grahamstown, and Colesberg.

The Eastern system of Railways has East London as its starting point, and runs through Fort Jackson, Blaney, Kei Road, Toise River, Queenstown, Sterkstroom, Molteno, and Burghersdorp to Aliwal North, 280 miles, with a branch line to King Williamstown. This line has been extended to join the Midland Railway (to Johannesburg) in the Orange Free State.

There are also railways belonging to private companies between Port Nolloth and the Cape and Namaqua United Copper Mines, 300 miles in length, and from Port Alfred to Grahamstown, 45 miles.

NATAL.—The Natal Railway system consists of a main line from Durban through Richmond Road, Pietermaritzburg, Howick and Ladysmith, and Biggarsburg to Charlestown, with a branch to the Dundee Coalfields, and has recently been extended to Johannesburg. There are also short lines from the Point, where passengers land, to Durban, a distance of two miles; and from Durban along the coast to Verulam, 19 miles, and Isipingo, 11 miles.

DELAGOA BAY.—A line, 129 miles in length, has been made from this Port to Nelspruit, and is being continued towards Pretoria and Johannesburg. A branch line is also being constructed towards the Murchison Gold Fields.

BEIRA.—From Fontesvilla, 40 miles by water from Beira, a line 118 miles in length has been opened to near Chimoio, and is being extended towards Fort Salisbury, 180 miles further.

LUGGAGE BY RAIL AND COACH.—Passengers by the South African Railways are allowed 100 lbs. first class; 50 lbs. second class; and 25 lbs. third class, free per adult. Children between 12 and 3 pay half fare and are allowed half the above quantities. Excess baggage is charged ½d. per lb. for distances up to 25 miles, ¾d. per lb. for distances between 25 and 50 miles, and ¾d. per lb. for distances between 50 and 100 miles, with ¼d. additional for every 100 miles, or portion thereof, beyond 100 miles. From Capetown to Kimberley the rate is, therefore, 2½d. per lb., and from Durban to Ladysmith 1d. per lb.

FARES.—The fares by the South African Railways are, with certain exceptions, 3d. first class. 2d. second class, and 1d. third class per mile. Return tickets are issued at a fare and a half.

Passengers preferring to pay their railway fares in London, can obtain tickets from Capetown, Natal, or Delagoa Bay, to the various inland stations, at Messrs. Donald Currie & Co.'s London Offices.

OTHER MEANS OF COMMUNICATION.—There is regular passenger communication by means of mail carts, coaches, and in some cases bullock waggons, between the railway stations and the larger South African towns situate at a distance from the railway lines.

The table following will show approximately the distances, by the various routes, to the South African Goldfields. The fares by coach vary considerably from time to time. Besides the coaches, much cheaper means of travelling are afforded by waggons, the fare by which, from either Kimberley, Vryburg, Viljoen's Dritt, Ladysmith, or Biggarsburg to the Goldfields, is usually only about £2 or £3.

BLOEMFONTEIN AND KIMBERLEY ROUTES.				NATAL ROUTE.		
Capetown to Kimberley by rail	...	647	Miles.	Durban to Charlestown by rail ...	...	303 Miles.
Capetown to Bloemfontein		749	„	Durban to Johannesburg by rail or coach	433	„
Capetown to Vryburg by rail		774	„	Durban to Pretoria by coach	...	463 „
Capetown to Mafeking by rail	...	870	„	**DELAGOA BAY ROUTE.**		
Capetown to Johannesburg via Bloemfontein				Delagoa Bay to Crocodilport by rail	...	113 Miles.
by rail	...	...	1,013	„	Delagoa Bay to Pretoria	...	350 „
Capetown to Pretoria by rail	...	...	1,040	„	Delagoa Bay to Johannesburg	...	373 „
Algoa Bay to Kimberley by rail	...	485	„	**BEIRA ROUTE.**		
Algoa Bay to Johannesburg by rail		713	„	Beira to Fontesvilla by steamer		40 Miles.
East London to Johannesburg by rail		665	„	Fontesvilla to Chimoio	...	118 „
				Chimoio to Umtali ..		80 „
				Umtali to Salisbury	...	149 „

SUNNY SOUTH AFRICA.

I.—A LAND OF DIAMONDS AND GOLD.

Africa, a land of surprises—South Africa, a land of diamonds and gold—The "Cinderella" of the Empire—Effects of the discovery of diamonds—The Gold Fields of South Africa—Ancient Workings—Ophir—Fluctuation of value of the diamond—Unlimited demand for gold—Effects of the discovery of gold in Australia—The "rush" in South Africa—Rapid rise of Johannesburg—Extraordinary richness of the Rand.

"*Semper aliquid novi Africa affert*"—so wrote Pliny nearly two thousand years ago, and the statement is as true now as it was then, if not indeed truer. Africa has always been, and still is, a land of unexpected discoveries and startling surprises. In the heart of Inner Africa, for instance, instead of arid deserts and broad savannahs like those to the north, or grassy plains and treeless uplands like those to the south, Stanley found a huge forest, over three hundred thousand square miles in extent, crammed with gigantic trees, so close that their branches interlaced one another, and formed an umbrageous canopy absolutely impenetrable to sunshine. Other parts of the African wonderland tell the same tale, and have given us undreamt-of solutions to many a geographical problem or sudden revelations of long-hidden sources of wealth. In the far south, we have seen "a land apparently destitute of resources, barely able to support its scanty population, living the most frugal lives, suddenly transformed, 'as by the stroke of an enchanter's wand,' into a perfect Sinbad's Cave of precious stones and gold." Southern Africa, especially, is indeed a land of surprises, but, as a recent writer remarks, it is difficult to imagine that any more startling surprises can be in store for us than have been witnessed within the last quarter of a century, during which "a desolate corner of a distant desert, shut out by barren wastes from communication with the sea and with the fertile districts of the country, has been converted into a teeming hive of industry" and an inexhaustible source of wealth. That, however, was but the "overture" to the "grand march" of South African progress !

Although pastoral, and, to a very limited extent, agricultural, industry had from the outset laid a broad and permanent foundation for real, if slow, progress, yet after a century and a half of apathetic Dutch dominance, and half a century of more eventful and progressive British rule, to stay-at-home Englishmen generally, the Cape remained almost as much a *terra incognita* as the interior of the continent. The country seemed to call for no particular notice, and, fifty years ago, was probably less known and talked about than any other considerable portion of our over-sea possessions. In truth, while her Australasian and Canadian sisters had got on in the world, had been gay and prosperous, and had received much company in the shape of emigrants, this "Cinderella" of the empire stood by her "Stormy Cape" neglected and almost ignored. Now and then a lurid light was cast across this far-off corner of the Dark Continent ; a massacre of settlers in some outlying district by the savage natives, and sharp reprisals by colonial or imperial troops awakened a strong but transient interest in "The Cape," but it served to repel rather than attract either the capital or the labour of the mother-country.

The *discovery of diamonds* in Griqualand West altered all this, and almost immediately produced a marvellous change in the condition and prospects of the country, which was, as it were, uplifted in a day from obscurity into universal notice, while its destiny was advanced hundreds of years at a bound. Mr. Reunert, in his excellent work, "Diamonds and Gold in South Africa"—a work which everyone interested in the country and its development should read—asserts, and that rightly, that during the four centuries which have elapsed since the Portuguese sailors, steering south in search of the sea-route to India, first sighted the Cape of Good Hope, no more important event has happened in South Africa than the discovery of the first diamond by Mr. John O'Reilly, in the month of March, 1867. "The beneficial effects of that discovery are apparent to-day in every corner of South Africa. It has spread new life and energy through all the Colonies and States, which, a quarter of a century ago, were in a languishing and impoverished condition ; and has converted the most despised possession of Britain into a source of wealth to the mother-country, and a field of ever-widening enterprise for her sons."

During the first five years after this auspicious discovery (1867-72) nearly two and a quarter million pounds' worth of diamonds was exported from the Cape ports. The output then gradually increased to 7¾ millions in 1873-77, 16¼ millions in 1878-82, nearly 16 millions in 1883-87, and 20½ millions in 1888-92. Altogether, the South African "Diamond Fields" have produced, up to the present, about 70 million pounds' worth of the gem ; and, as Mr. Reunert points out, of this enormous sum realised by their sale, probably one-half has been paid away in wages at the mines, and for other local expenses. As a natural result, the trade of the country largely increased ; other industries revived ; public works were energetically pushed forward ; means of communication improved and extended ; natives were employed in ever-increasing numbers, and "taught to work instead of to fight" ; while exploration and settlement steadily advanced north to, and beyond, the Zambesi.

But wonderful and widely-beneficial as the results of the diamond discovery have been, and still continue to

be, they are completely overshadowed by the more recent discovery of enormously richer and equally inexhaustible sources of wealth. "The Gold Fields of South Africa," though as yet in their infancy, already rival those of Russia, Australasia, and California ; and the output of the precious metal, especially from the Witwatersrand mines in the Transvaal, is steadily increasing, and at such a rate that South Africa, which now ranks *fourth* among the gold-producing countries of the world, must soon rank first. Russia and Australasia now produce yearly about 6 million pounds' worth of gold each, or about half a million less than the present annual output of the United States. Witwatersrand alone, *from a narrow strip of country not more than ten or twelve square miles in area*, constituting only the first row of claims on the outcrop of the Main Reef series, already produces over 5 million pounds' worth of gold per annum. Besides the Rand, a number of other gold areas in the Transvaal are being more or less actively worked, and the recently-opened fields in the Witwatersberg, some 30 miles north of Johannesburg, may rival those of the Rand itself. In fact, the whole of the eastern portion of the Transvaal may be regarded as *one continuous gold field*. Native gold is also known to exist in several parts of the Cape Colony, and may yet be found in paying quantities. Careful prospecting operations in the Orange Free State have resulted in the discovery of permanent and payable reefs, while the "banket" beds of Vryheid, in Dutch Zululand, are identically similar in formation to those of the Witwatersrand, and probably of equal richness. Swaziland, Mashonaland, and Matabeleland are also rich in gold, the gold-bearing area extending far north to the Zambesi. In fact, we may say that the entire country, from the great bend of the Limpopo to the south end of Lake Tanganyika, must be gold-bearing, as it is in all directions honey-combed with "Old Workings," and is by some supposed to be the *Ophir*, whence King Solomon is said to have drawn gold to the value of £900,000,000 sterling.

A well-known mining expert, Mr. Robert Williams, speaking at a banquet given on St. Andrew's Day, 1892, at Salisbury, the capital of Mashonaland, stated that he had travelled some 4,000 miles over almost continuous old workings,[*] and he estimated that at least 800,000 tons of ore had been excavated by the old diggers. These ancient

---

[*] "These old workings," says Mr. Selous, "are of a very singular and persistent character, consisting for the most part of circular shafts, varying in depth from 20 to 80 feet, but not more than 30 to 36 inches in diameter. They have been sunk at all sorts of distances apart, in many cases not more than one foot, and in others as much as fifty or a hundred feet. No outcrop is apparent on the surface, and nothing at the bottom of the shafts would seem to suggest a likelier reason for the stoppage of work than the gradual deterioration in the grade and size of the veins."

A curious fact in connection with these old gold workings is also mentioned by Selous, and that is—wherever lemon trees grow, old workings will invariably be found in the neighbourhood. The natives have no tradition as to how these trees have been introduced, and Selous thinks they may have been introduced by the Portuguese, two or three centuries ago, or they may date back to much more ancient times, when South-East Africa was visited by the trading peoples of Asia and Arabia in search of gold.

---

miners, however, seem to have given their attention exclusively to high grade ore, being doubtless unable to deal with low grade and refractory ores. They also seem, says Mr. Fairbridge, in his report on the Mashonaland Goldfields,[*] to have preferred open cuttings to subterraneous tunnelling, and apparently in few cases did their probably rude appliances permit them to go deeper than a hundred feet. "As might be supposed, the *débris* thrown out of their workings were a means of calling the attention of travellers, as they have later guided the prospector, to the existence of gold-bearing reefs in this country. But as the ancients were unable to exhaust the veins they struck, so also were they unable to complete their discoveries of good surface outcrops. On every field, since the arrival of the whites, excellent, and frequently very high grade, lodes have been struck, bearing no vestige of human prospecting or labour. Mashonalanders, as the new colonists call themselves, perhaps justly claim as a sign of the pre-eminent richness of their mineral rock, that in Zambesia alone of old-world places have the old Eastern nations thought it worth while to delve upon a gigantic scale for the commodity whose value has been as old and long-established almost as the hills themselves."

South Africa, then, is pre-eminently *a land of diamonds and gold*.

"The stones thereof are the place of sapphires, And it hath dust of gold."[†]

Both the gem and the metal are alike wonderful energisers of trade and industry, but there is this important difference between them. The "demand" for diamonds—as for coal, iron, copper, and even silver—has to be taken into account, and necessarily regulates the production and the price. Over-production would be the bane of the diamond, as of the baser mineral and metal industries ; and, as a matter of fact, although the great diamond corporation—the De Beers Consolidated Mines, Limited—which produces over ninety per cent. of all the diamonds mined in South Africa, and exercises a paramount control over the industry, paid over three millions sterling for the greater portion of the Dutoitspan and Bultfontein mines, no work has been done by the Company at either of these mines since 1892, as a sufficient supply of blue ground is more readily and profitably obtained from the two principal mines—De Beers and Kimberley. But there are certain risks, more particularly—as Mr. Rhodes pointed out in his speech at the annual meeting of the De Beers Company, in 1893—the risk of new mines being suddenly discovered, and worked recklessly, to the detriment of the industry generally. In that case, the diamonds would have to be sold for what could be got for them, perhaps for considerably less than the cost of production, and this, of course, would speedily lead to the annihilation of the industry. Conducted, however, systematically on scientific principles, and under wise and vigilant control as at present, the life of the diamond industry is practically unlimited—there are plenty more 'pebbles' in the ground, and plenty more on the floors. And in order to control still more

---

[*] Appendix XV. in Mr. Reunert's book, "Diamonds and Gold in South Africa."

[†] Job, chap. xxviii., verse 6.

effectively the output and price of what is essentially an article of luxury, the Consolidated Mines Co. have bought up large areas of land around the Kimberley mines, and have acquired a third of the land in British Bechuanaland, and a preferent right to any diamonds that may be found in any part of the enormous territories of the British South Africa Co.

Gold, on the other hand, is an article whose standard of value will not be changed or affected in the slightest by any probable or possible extension of production. As Mr. Hamilton Smith points out, gold is now the only material for which there is a practically unlimited demand, and as over-production is therefore an impossibility, the richest and most extensive gold-bearing areas in the world, as those of South Africa are, will certainly be still more vigorously worked and extensively developed, with the same beneficial and enduring results as in Australia. Fifty years ago, Australia was a country little known to the mass of the people at home ; but when at length, in every bookseller's shop in Great Britain, maps of Victoria appeared, dotted over with yellow marks, showing that gold had been discovered in all directions, there was a mighty "rush" to the land where wealth unbounded was to be obtained. In the colony itself, the entire population became "drunk with gold." Settlers left their homesteads, professional men their offices, sailors their ships, and rushed off to the "diggings." For a time there was an excitement which nothing could allay, and it was not until the hardships and dangers of a digger's life, with its uncertain results, began to show that none but the strong and experienced could succeed, that the country returned to its normal state, and gold-mining became a regular and steady industry. The effects of the discovery of gold in such abundance proved, however, to be as permanent as they were startling. By its magic touch, tents were transformed into flourishing villages and mud huts into magnificent cities. Little thought the three solitary pioneer settlers of 1835, when they built their mud huts on the then dismal banks of the Yarra, and surveyed the desolate wastes around, that in fifty years a colossal city would cover them, or that the dreary spot, then bought from the natives for two blankets and a bottle of spirits, would be the site of, with one exception, the most populous city in the Southern Hemisphere, and the most important commercial centre in our Australasian empire.

So much for the colonising power of raw gold in Australia—a power by which a similar transformation will be seen in the far interior of South Africa. Here, indeed, a like process was begun in 1867, when the diamond was discovered, with the result that "The Camp" of the early diggers, with its motley collection of tents and tin houses, became a well-built town, furnished with all the necessities and luxuries of civilisation. Gold, however, did not assert its power in South Africa until about ten years ago, and although the "rush" to the South African Goldfields has never attained anything like the proportions of its Californian or Australian prototypes, yet discovery after discovery of the precious metal in various parts of the country gave rise to a gold mania, which speedily developed into mad speculation and gambling in

shares in properties that not only were, in most cases, not being worked, but only very roughly and superficially, if at all, tested—the inevitable result being disappointment and ruin to hundreds of too credulous investors, too eager to be "in the swim." At the outset, there was much feverish activity on the Exchange and in the Share Market, and miserably inadequate work on the reefs ; people, profoundly ignorant of their real value, dabbled in stocks "boomed" in glowing prospectuses, and exchanged their money for worthless scrip ; and thus, in a few years, millions were lost in all but fruitless speculations. Fortunately for the reputation of South Africa as a gold-producing country, the marvellously rich and practically inexhaustible conglomerate reefs (locally called "banket") of the Witwatersrand were discovered, and soon attracted both capital and labour in abundance, with the result that, on the highest ridge of the High Veldt of the Transvaal, one thousand miles from Cape Town, we find, instead of the few miserable tents and shanties which formed the Johannesburg of 1886, a large town, solidly built, with macadamised roads and broad streets, lighted by gas and electricity, stores filled with the newest goods and the most modern mining appliances, shops stocked with the latest fashions, tramways from one end of the city to the other, a distance of over three miles, numerous suburbs, and an ever-increasing number of outlying townships along the Rand. The latter are connected with Johannesburg itself by a light railway, which, at Elandsfontein, seven miles distant, is crossed by the main trunk-line from the Cape to Pretoria, the objective also of the East Coast railways from Delagoa Bay and Natal, the former of which is already open for some distance within, and the latter to, the Transvaal frontier.

Before the discovery of diamonds, *wool* was by far the most valuable item in the export trade of South Africa. Then, until very recently, the *diamond* took the first place, but now both diamonds and wool are eclipsed by the *gold* output. The production of wool and diamonds for some years past has remained almost stationary, the former averaging a little over two millions sterling, and the latter about four millions yearly. The output of gold, on the contrary, has advanced by leaps and bounds, and now amounts to about five millions a year ; and, with plants of increased capacity, the Rand "banket" beds alone are expected, in a very few years, to yield ten million pounds' worth per annum, if not considerably more. The value of the few square miles included in the Rand Goldfield is incredible, and two well-known mining engineers—Theodore Reunert and Hamilton Smith—have estimated the total quantity which it may be expected to yield, and, though the methods of calculation were different, the results arrived at are much the same. Mr. Reunert points out that the Johannesburg Main Reef Series have been exposed along the outcrop for at least 30 miles, and assumes that they will be within reach of mining operations for probably several miles across the dip. Fixing the limit at only one mile, there are 30 square miles of auriferous beds. Of the two or three hundred feet which these beds measure in thickness, Mr. Reunert allows only five feet as carrying gold in payable quantities.

According to the last returns of the Witswatersrand Chamber of Mines, the average yield of the district is at present 10 dwts. of gold per ton crushed.*

It is known that a good deal of gold is lost, which more perfect treatment will enable to be saved in the future; but, taking a low estimate, Mr. Reunert assumes those five feet of payable "banket" to carry an extractible average of only 8 dwts. per ton. Thus he arrives at a total of 130 million ozs. of gold, worth, say, 450 millions sterling, "as the value of the ore locked in Nature's treasury, and only waiting the industry of man for its extraction." Mr. Hamilton Smith estimates, in the stretch of 11 miles, a paying length of 50,000 feet, a probable thickness of fully 5 feet, and an inclined depth of 5,200 feet. This aggregates 100 million tons, of which 3 million tons have been mined; the remaining 97 million tons, at an average of 12½ dwts. to the ton, would yield 60 million ozs., having a gold value of £215 millions. The many miles of "banket," outside this district of 11 linear miles, will yield at least one half of this amount, or, say, £325 millions in all. "This seems a huge figure, but it is by no means a wild conjecture, and the final results will probably exceed this sum." These estimates have

* "This is merely the yield from the mill; but subsequent treatment, by cyanide and other processes, brings up the total average yield to over 12½ dwts. per ton."—*Reunert.*

been further confirmed by the successful results shown by the "Great Borehole" of the Rand-Victoria Mines on the Boksburg line. Not less than 2,343 feet of barren strata were penetrated by the drill, but at that depth the *South Reef* was struck, and, 54 feet lower, the *Main Reef* itself was pierced, the footwall being struck at a depth of 2,401 feet, and giving an average assay through the 4 feet of 1 oz. 15 dwts. The Main Reef Leader, 18 inches wide, was struck at 2,391 feet, and assayed in some samples as high as 10 ozs. per ton, showing visible gold. The *Johannesburg Star* estimates that the Rand-Victoria Mines have, at a moderate computation, 12 million tons of ore, computed to give an average result of 23 dwts. to the ton. "Take, as with our experience of banket we fairly may, this as a fair specimen of the results to be obtained from the Nigel to Rand-fontein, and the fabled Eldorado of Sir Walter Raleigh is but pinchbeck. Then, when we consider that, at this enormous depth of 2,397 feet, we have not to deal with a densely-pyritic ore, in which the precious metal is enveloped in a non-amalgamating cover, but with free gold in considerable quantity, the troublous problems of ore treatment sink into comparative insignificance, and leave us a clear vista, wherein will be an output enhanced maybe ten times, and whereon not only our own but the eyes of the whole world will gaze with appreciation and envy."

II.—THE COUNTRY OF THE VELDT AND THE KARROO.

The Veldt—A sea of grass—The Karroo—Latent fertility—The coasts of South Africa—Walfish Bay—Table Bay and Cape Town—Simon's Bay and Simonstown—Cape Agulhas—Mossel Bay—Algoa Bay and Port Elizabeth—Port Alfred—East London—St. John's River—Port Natal and Durban—Delagoa Bay and Lorenzo Marquez—Inhambane—Chiloane—Sofala—Beira—Chinde—Quillimane—Mozambique—Physical aspects of the country—Gradual rise in terraces from the seaboard—The Coast Plateau - The Southern Karroo—The Eastern Uplands—The Northern Karroo—The rivers and lakes of South Africa—"Bars" at river mouths—Falls and rapids—South Africa, once a country of great lakes and vast inland seas—Vleis and salt-pans—Political divisions and governments.

The "Veldt" and the "Karroo" are the peculiar and distinctive features of South Africa. There is nothing English to which they can be compared; in fact, there is nothing in the Empire like them—the nearest resemblance to the true "Veldt" and the monotonous "Karroo" being, perhaps, the downs and the waterless plains of Queensland.

The Dutch word "veldt," like the German "feld," is literally synonymous with our word "field," but, in South Africa, the term "veldt" has a much wider application than its English or German equivalents, as it is given not only to the wide, rolling pasture-lands, covered by rough scrubby grass, or by more or less dense growths of mimosa or acacia and other bushes and scrubs, but also to the herbage itself, which is spoken of as "sweet veldt" or "sour veldt," as the case may be; while, according to the season, the farmer moves his flocks and herds from the "hooge veldt" to the "bush veldt," or from the "koud veldt" to the "warm veldt." In the Transvaal, the higher portion of the plateau is known as the "Hooge Veldt," the hilly country to the north and east being distinguished as the "Banken Veldt" or terrace country; the sub-tropical, tsetse-infested bush country along the

Limpopo being the dreaded "Bush Veldt" of the Transvaal Boer. In the Cape Colony, we have the well-known plateau of the "Warm Bokkeveldt," and further north the higher and more exposed "Cold Bokkeveldt." In one form or another, the term "veldt" is of universal application throughout South Africa.

During the rainy season, the South African veldt is a "sea of grass," affording abundant pasture to millions of sheep and cattle, and to the antelopes and other ruminants which are still found in considerable numbers in the more inaccessible parts even of long-settled districts. But "when the strength of the African sun is at its greatest in summer, the veldt is very hot and barren-looking, its brown and parched surface cracking into large fissures or 'sluits,' as they are locally called, and radiating back the light with a strange simmering mirage, deceiving the eye and perplexing the judgment of the stranger." So clear is the atmosphere that distances are dwarfed, and mountains that are miles away appear quite close. "The roads across the veldt are not macadamised, but wind away as tracks cut out according to the whims and fancies of a post-cart driver, and twisting in long and sinuous lanes over the interminable spaces. Over these the slowly-

moving ox-waggon, with its white canvas covering, makes its way from point to point, and across 'drifts' or fords of the rivers."* The halting-place, or "outspan," is a piece of land reserved for the public use, and "here, in the summer noon, the kurveyor or waggoner is seen, with his unyoked span around him, taking his siesta, and preparing for the evening, or perhaps, if it is moonlight, the night journey."

The "Karroo" is another peculiar feature of South Africa. The name is taken from the little *karroo* plant, one of the best kinds of bush for ostriches as well as for sheep and goats, and is applied to a marvellous tract of country, about two-thirds the size of Scotland, in the interior of the Cape Colony, apparently an arid desert, but possessing an extraordinary latent fertility, and requiring only sufficient moisture to be as productive as the banks of the Nile. The name "karroo" is also given to similar tracts both to the north and south of the Karroo proper, or the Great Karroo ; in fact, all South African plains and plateaux, which are, as it were, intermediate between the grass- or bush-covered veldt and absolute desert, are karroos. During the long periods of drought, the parched karroo is devoid of verdure, but, when rain falls, the ground is quickly clothed with grass and shrubs, and parts of it have the semblance of a brilliant flower-garden. When thunder-clouds break over any area, and heavy rain falls, it is marvellous to see the magical transformation of the sparsely-covered soil ; grasses and flowers springing up with great rapidity and in countless variety, carpeting the surface with every colour and hue"—one of the many startling contrasts and sudden surprises which excite the interest and admiration of the sojourner from other lands in which Nature is not, so to speak, so impulsive and erratic in her movements.

South Africa is then, *par excellence*, the country of the Veldt and the Karroo. Life on the veldt is full of interest and enjoyment ; mere existence is a pleasure ; while, unattractive as the karroo may appear, "its sunny sky, its translucent atmosphere, its dry buoyant air—'exhilarating as wine to the senses'—its starry, balmy, and dewless nights ; its measureless expanse ; its vast and unbroken solitude ; and even its weird desolateness, have a peculiar charm, which clings to the memory of those who have dwelt on any part of it."

The *coast* of South Africa is of the same regular and unbroken character as that of the continent generally, being "singularly deficient in good harbours, devoid of navigable rivers, and washed a great part of the year by a most tempestuous ocean, girdled by a never-ceasing surf, while its projecting capes and headlands bristle with reefs, on which many a gallant ship has met its fate."† Nearly the whole of the western coast, from the mouth of the Cunene to that of the Orange River, is low and sandy, and the adjoining coastland is barren and dismal, with no permanent rivers, and scarcely any vegetation. *Walfish Bay*, the only point of any importance on this dreary coast, belongs to the Cape Colony, and may derive

some importance in the near future as the starting-point of a railway into the interior ; meantime it serves as the port of entry for supplies to the German officials at Windhoek, the capital of German South-West Africa, a small settlement in the Damara highlands, some 200 miles inland. Thence to the Orange River, not a single perennial stream enters the sea, and not a vestige of human occupation is visible except at one or two small bays and anchorages, such as *Sandwich Harbour*, where large quantities of fish are caught and cured, and *Angra Pequeña*, or Lüderitz Bay, whence a good cattle-road leads into the more habitable and inviting uplands of Great Namaqualand. The *Orange River*, although it is 1,200 miles in length, is a terribly disappointing stream, and the country it flows through, in the lower part of its course, is about the most dismal and barren in the world. The river itself is a finer stream hundreds of miles inland than it is at its mouth, and, moreover, an impassable bar forbids entrance from the sea, except for a few days after very heavy rains in the interior. Usually, however, inside the bar, it can be ascended by small craft for some 30 miles. We then come to *Port Nolloth*, the coast terminus of a light railway from the copper mines at O'okiep. Thence to the broad curve of *St. Helena Bay*, the coast is still low and desolate, with one or two lonely harbours and a few river mouths hopelessly blocked with sand or rocks. Rounding Cape St. Martin, and passing through a narrow entrance, we enter a splendid land-locked basin, *Saldanha Bay*, one of the finest natural harbours in the world ; a noble bay, easy of access in all winds, but seldom visited.

Passing a few rocky islands, we enter *Table Bay*, along the curving shores of which extends the metropolis of all South Africa—its main avenues running straight from the sea to the pine and silver-tree plantations, which clothe the base of a titanic wall of rock, the flat-topped Tafel Berg or Table Mountain, that, with the picturesque Devil's Peak and the grotesque Lion's Head on either flank, enclose the city and its immediate environs, forming an amphitheatre comparable in scenic effect to Naples or Rio.

Cape Town is not only the capital of Cape Colony, but, with its 60,000 inhabitants, is also the most populous town, and, with the exception, perhaps, of Port Elizabeth and Johannesburg, the most important commercial centre in South Africa.

Originally laid out with mathematical precision by its stolid and unimaginative Dutch founders, the main streets run parallel to each other, and are crossed at regular intervals by minor streets. At the top of the chief thoroughfare—Adderley Street, which compares favourably with the main street of any ordinary English town—is a magnificent avenue of oaks, planted by the early Dutch settlers, and still the favourite promenade, leading to the official residence of the Governor and High Commissioner. The town contains several fine public buildings, the finest not only in Cape Town, but in all South Africa, being the new Houses of Parliament, completed in 1886 at a cost of a quarter of a million sterling. The most interesting building, however, is the Public Library, with its priceless literary treasures presented by a former Governor, Sir George Grey, and the Museum, containing

---

* Greswell's *Africa, South of the Zambesi*. (London : Stanford).

† S. W. Silver's *Handbook to South Africa*, p. 553.

specimens of almost every species of South African antelopes and birds. Rough diamonds in the matrix, gold nuggets and quartz may also be seen here. In the magnificent Botanical Gardens, experiments are being constantly carried on with plants and trees from other countries, to test their suitability for the soil and climate of South Africa. As a port, Cape Town ranks first in South Africa, and by the completion of the great breakwater and commodious docks, the Bay has been converted from a dangerous roadstead into a safe and convenient harbour, accessible at all states of the tides, and completely protected from the fearful effects of the northern and north-western gales, by which many a vessel has been driven ashore. The Castle is a quaint specimen of the old Dutch citadel, but it is perfectly useless from a military point of view, being on all sides commanded by the adjoining hills.

Strong batteries, erected along the foreshore and on Signal Hill, and garrisoned by imperial troops, defend our " Half-way House " to India and the East from hostile attack, while the town is connected by rail with the naval station at Simon's Bay on the other side of the peninsula. The suburbs and environs of Cape Town are exceedingly beautiful, and by almost any of the roads or routes from the city magnificent ocean and mountain views may be enjoyed, and the glowing heat and dusty streets may be quickly exchanged for cool health-giving breezes in the most delightful sylvan retreats. *Wynberg*, a pretty village on the eastern side of Table Mountain, is usually regarded as the suburban limit.

The glory of Cape Town is its magnificent mountain, which " rises behind the town in a sheer precipice to the height of nearly 4,000 feet, cutting the sky-line with a jagged horizontal front nearly two miles in length." The ascent of the mountain is easily accomplished, the only danger being the dense clouds that suddenly collect and envelop the summit, forming what is locally known as the " table-cloth." There are many other table or flat-topped mountains in South Africa, but the Cape people boast that "there is but one Table Mountain ;" and, indeed, " only those who explore the mountain can form any idea of the beauties hidden among its rocks. The frowning precipices which, seen from a distance, speak only of the convulsions of nature, are found on nearer approach to open into tiny glens and valleys, adorned with streams and cascades, and clothed with the most beautiful foliage and flowers. The flat summit of the Kasteel-Berg, or Castle Mount, which forms the buttress of the great precipice overlooking the Bay, is a miniature continent in itself, its surface diversified by river and hill, and producing a flora to be found nowhere else."[*] The view from the summit is magnificent, as also are the views to be obtained from the Devil's Peak (3,300 feet) and the Lion's Head (2,000 feet), which complete the majestic rock-wall that forms the amphitheatre in which Cape Town is situate.

The Cape Peninsula terminates in the real " Cape of Good Hope "—the " southern point of Afric's coast " immortalised by the great Portuguese poet, Camoens, in his " Lusiad " :—

> " I am that hidden mighty head of land
> The Cape of Tempests fitly named by you,
> Which Ptolemy, Mela, Strabo never fand,
> Nor Pliny dreamt of, nor old sages knew.
> Here in South Ocean end I Afric's strand ! "

From the lighthouse on *Cape Point*, as it is locally termed, the visitor obtains a fine view of the waters of False Bay and the broad expanse of the South Atlantic ; the Cape itself, a lofty sandstone precipice nine hundred feet in height, is certainly a far more striking object on this, the most picturesque and grandly beautiful part of the South African coast, than the low shelving bank of *Cape Agulhas*, which forms the geographical extremity of the continent, and from which also is drawn the theoretical boundary between the Atlantic and the Indian Ocean.

Within *False Bay* is the safe and well-sheltered *Simon's Bay*, on which stands *Simonstown*, the strongly-fortified station of our fleet in South African waters. About midway between the Cape of Good Hope and Cape Agulhas is *Danger Point*, near which H.M.S. Birkenhead struck in 1852—an ever memorable disaster, in which British soldiers exhibited a calm courage infinitely more heroic than was ever displayed in the most desperate charge on the battlefield. Cape Agulhas, or the Needles, is so named from the sunken rocks or saw-edged reefs which run far out to sea, and with the strong currents and furious storms met with in the channel between the coast and the outlying *Agulhas Bank*, render its navigation difficult and dangerous.

It is off this justly-dreaded point that the great Mozambique, or, as it is sometimes termed, the Agulhas current, which sweeps down the Mozambique Channel and follows the curve of the South African coast, bringing with it the warm waters of the Indian Ocean, meets the Antarctic or Cape current, another powerful current that flows from the cold waters of the South Polar seas[*]—hence the continual tempests and dangerous navigation, these currents being but too often, in thick weather, the unsuspected cause of many wrecks.[†] So thoroughly, however, are these seas known to the officers of the regular ocean and coasting steamers, and so admirably do they handle their vessels, that, practically, there is now as entire an immunity from disaster as on any of the great ocean routes.

The next noteworthy point on the coast is *Mossel Bay* —a port of call for the coasting and intermediate steamers, with an excellent harbour protected from westerly gales by Point St. Blaize, and situate about halfway between Cape Town and its energetic and successful rival, Port Elizabeth. We must not, however, pass on without noticing the curious chain of lakes near the margin of the sea about five miles from Woodville ; and, further east, the picturesque land-locked estuary of the *Knysna*, a splendid natural harbour, entered by a narrow passage between lofty sandstone cliffs, and with no less than 14

* Brown's *South Africa*, a practical and complete Guide for the use of Tourists, Sportsmen, Invalids, and Settlers. (London : Sampson Low & Co.).

* Greswell.   † Silver's Handbook.

fect of water on the bar. The coasting steamers call here regularly, and the harbour could easily be made impregnable and absolutely safe from any hostile attack from the sea.

Passing by the shallow indentations of *Plettenburg Bay* and *St. Francis Bay*—the coast between which is exceptionally dangerous, and has been the scene of numerous wrecks, we double the low rocky point of *Cape Recife* and enter the broad and well-known inlet of *Algoa Bay*, on the north-western angle of which stands *Port Elizabeth*, the "Liverpool of South Africa," and the most important centre of the foreign trade of the country. Even before the construction of the railway, which now runs from the port to the Free State and the Transvaal, connecting at Middelburg with the East London line and at De Aar with the Cape trunk line, Port Elizabeth was a stirring and bustling trading-place, "especially during the wool season, when the huge transport waggons, carrying from 6,000 to 10,000 lbs., came in laden with bales of wool, skins, and ivory, to load up again with merchandise for the interior towns and villages." The anchorage is sheltered from the winterly north-west winds, but is exposed to the heavy rollers caused by the frequent easterly gales. There are two wrought-iron jetties ; but passengers and cargo have to be landed by means of tugs and lighters. The European population now numbers about 15,000, mainly the descendants of the energetic and enterprising British settlers who founded the town in 1820, and occupied the then unsettled "hinterland."

The eastern coast of the Cape Colony, though, on the whole, as regular and unbroken as its north-western seaboard, is infinitely more beautiful and attractive. The evergreen slopes, picturesque bays, and wooded kloofs which diversify this coast, form a pleasing contrast to the low, sandy, barren, and desert shore-line on the opposite side of the colony. About 100 miles to the eastward of Port Elizabeth is *Port Alfred*, beautifully situate at the mouth of the Kowie River, and justly named the "Dartmouth" of South Africa—a name also claimed by the more important port of *East London*, at the mouth of the Buffalo River, which flows through equally picturesque scenery. Port Alfred, or the Kowie, as it is also called, is the outport of *Graham's Town*, the metropolis of the eastern division of the Cape.

East London was originally a mere outpost to *King William's Town*, the capital of the formerly separate colony of British Kaffraria ; but, since the extension of the Eastern Railway System, of which it is the coast terminus, and its junction with the Midland and Western Systems, it has become the centre of a very large and rapidly increasing trade, second only to that of Port Elizabeth and Cape Town. Extensive and costly harbour works, planned by Sir John Coode, have so far removed the obstructions at the mouth of the Buffalo, that steamers of 4,000 tons can enter the sheltered anchorage inside the bar.

Thence, for three hundred miles, the only serviceable harbour is that of *St. John's River*, on the Pondo coast. The River St. John, or Umzimvubu, enters the sea between two huge forest-clad cliffs, which, with the surrounding forests and many-coloured cliffs, are among the most

romantic and striking scenes on the South African seaboard—a scene, however, equalled, if not surpassed, by the *Port of Natal*, with its tall "Bluff" overlooking a broad bay, along the northern shores of which extends the prosperous and thriving town of *Durban*, backed by the wooded plateau of the Berea, now covered with handsome mansions and pretty villas. Durban is not only the seaport, but also the most populous town and the principal trade centre in Natal ; and from the "Point," the Natal main line of railway runs right through the colony to Charlestown, on the Transvaal border, whence it will be extended to Johannesburg. Two short coast-lines also run from Durban —one to Verulam, 19 miles to the north-east, and the other to Isipingo, 11 miles to the south-west.

Few indeed and small are the vessels that have any dealings with the Zulu or the Tonga coasts. Both are bordered by shallow tidal lagoons, and even a brief stay in these hot and marshy coastlands is sure to bring unpleasant consequences in the shape of fever and ague. This has also been a drawback of the great inlet of *Delagoa Bay*, one of the finest natural harbours in the world, spacious, deep, and well sheltered, where the largest ocean-going vessels can lie in perfect safety. The town of *Lorenzo Marquez* has been notorious for its unhealthiness during the rainy season, but its increasing importance as the terminus of the shortest route from Pretoria and Witwatersrand to the sea, will, no doubt, be followed by improved sanitation and healthier conditions. From the port a railway runs to Komati Poort, at the foot of the Lobombo Mountains, whence it is being extended to Pretoria *via* Middelburg, the total distance being 350 miles, as against 1,000 miles from Cape Town, and 500 miles from Durban. The main line is now open to Nelspruit, 129 miles from Delagoa Bay, and branch lines are being made from Komati Poort, towards the Murchison Gold Field, on the north, and from Crocodilpoort, 113 miles from Delagoa Bay, to Barberton, in the De Kaap Gold Field.

From Delagoa Bay the coast trends north-east, and sailing in that direction, we pass the mouth of the Limpopo river, which, though shallow, is navigable by sternwheel steamers, and, doubling the well-known Cape Corrientes, we reach *Inhambane*, a sleepy old Portuguese town at the head of a deep bay, backed by wooded hills. Thence the coast turns north, curving at the delta-mouth of the Sabi river, past *Chiloane*, a port of call on a small island on the coast, towards the broad bay into which opens the estuary of the Sofala river, the fine natural harbour of the historically famous old town of *Sofala*, the oldest of all the Portuguese settlements in Eastern Africa. "Its name is that of a maritime kingdom, renowned in ancient times for its wealth, which formed part of the mythical empire of Monomotapa, of which the earlier travellers gave marvellous accounts. From its richness in gold and ivory, Sofala has been even supposed to be the golden Ophir to which King Solomon sent a fleet of ships every three years," and from which he is supposed to have obtained enormous quantities of gold. As we have already stated, the "Hinterland" of the Sofala coast is literally honeycombed with ancient workings, and the wonderful ruins of numerous "Zim-

labwes" all over the country are undoubtedly the remains of the strongly fortified stations and temples of the ancient miners, who probably invaded the country, and forced the inhabitants to labour for them much in the same way as, in later times, the Spaniards exploited the gold and silver mines of Mexico and Peru.

The riches of the golden "Hinterland" of Sofala will not, however, flow through their ancient outlet, but through the bustling brand-new little port of *Beira*, at the mouth of the Pungwe river, the estuary of which forms an excellent harbour, while the river itself is navigable for 40 miles inland to *Fontesvilla*, the present terminus of the Mashonaland railway, of which 75 miles (to a station near Chimoio) have been opened—the rest of the journey to Salisbury, 180 miles further, being made by waggon. The estuary and lower course of the adjoining Busi river are also navigable, and, like the Pungwe, might be utilised to some extent in facilitating passage and transport over the unhealthy flats of the littoral to the healthy and elevated plateaux of Manicaland and Mashonaland.

From Port Beira, a sail of 100 miles along a low and uninviting coast brings us to the delta of the Zambesi, the largest of all the African rivers that flow into the Indian Ocean. This great river is navigable for small steamers as far as Tete, about 260 miles inland; beyond that, a succession of cataracts and rapids limits navigation to a few sections of the river. The Anglo-Portuguese Convention of 1891 gave England a foothold at the *Chinde* mouth, the concession to be used for landing, storage, and transhipment of goods. At Chinde, says Dr. Rankin, the river is about a mile wide, and lower down it increases in width, whilst its surface is covered with islands. From the immense volume of water brought down, the banks and channels are continually undergoing change. The greater part of the delta is made up of gently-rolling grass country, the depressions of which during the rainy seasons are covered with water. The inundated districts at the seaboards, swamped at high tides and floods, are thickly covered with mangroves, and but slightly populated. The Chinde River is undoubtedly the best entrance to the Zambesi, and affords a waterway for craft of from 400 to 500 tons.

About 50 miles higher up the coast, the Quilimane river enters the sea. The estuary was formerly the most northerly delta branch of the Zambesi, the Mutu channel, which connected it with the head of the delta, being then large and navigable all the year round; but it has long since been blocked up by silt and vegetation, so that *Quilimane* is completely debarred from access to the Zambesi itself except by sea. But this port, though thus cut off from the great river, and surrounded by swamps and marshes—"a place of mud, fever, and mosquitoes"—is still of some importance; its trade is fairly good, while its beautiful mango groves, its shady avenues, and delicious oranges, are delightfully refreshing. Some miles higher up the coast, we arrive at a much busier port, situate on a coral islet close to the shore. This is *Mozambique*, the capital of all Portuguese East Africa, and, curiously enough, lying under almost exactly the same parallel as *Mossamedes*, on the west coast. An irregular line be-

tween these two places would mark the furthest limit, territorially and climatically, of Southern Africa. All beyond, and even in some parts within that line, belong absolutely to Tropical Africa, scarcely any portion of which is suitable for permanent European colonisation. Europeans may live there for years without breaking down, but they cannot settle and work there, as they can in almost every part of *Temperate South Africa*.

Having now made the circuit of the South African coasts, we proceed to note very briefly its mountains and plains, its rivers and lakes, supplementing our *résumé* of the physical features of the country with a short account of its political divisions—a matter which, from the number and variety of States, Colonies, and Protectorates, is rather confusing to the general reader.

Regarding South Africa as a whole, we may say that the greater portion of the country forms a vast upland, which slopes towards the seaboard not regularly, but by a series of successive terraces or steps, the more or less abrupt seaward edges of which are marked by long ranges of mountains and hills. These gird the country in irregular lines, separated from each other by valleys and plains, and intersected at intervals by deep ravines or gorges (kloofs)—the main elevations trending generally in a direction parallel to, and at no very considerable distance from, the coast.

"Africa," says Professor Drummond, "rises from its three environing oceans in three great tiers, first a coast line, low and (in Tropical Africa) deadly; further in, a plateau the height of the Grampians; further still, the higher plateaux, extending for thousands of miles, with mountains and valleys." The relief of South Africa exhibits the same terraciform aspect that characterises the build of the continent generally.

"If the traveller," says the Rev. W. P. Greswell, in his exhaustive "Geography of Africa, South of the Zambesi," "were to land at Mossel Bay on the south coast and journey northwards towards the interior, he would see, immediately facing him, a coast range up which he would find his way through the Montagu Pass into the George district. Here a second range of considerably higher mountains, called the Zwartebergen, would interpose a barrier to the north, through which he would penetrate by Meiring's Poort or pass. Here he would stand upon the plateau of the Great Karroo. Further north still, however, he would see a third range, called the Nieuwveld Mountains, past which he would go by Nel's Poort—a defile which the railway engineers have utilised for the Western Railway from Cape Town to Kimberley. A similar impression of the 'step-by-step' rise' of the land would be gained if the traveller, journeying by rail from Cape Town to Nel's Poort, notices the gradients as he proceeds up the line through the famous Hex River Pass, past Montagu Road, Prince Albert Road, and so on to Beaufort West and Nel's Poort." On the Natal side, the same formation would be clearly seen during a journey along the main line from Durban, through Maritzburg, Estcourt, and Ladysmith, and thence by the branch line over Van Reenen's Pass in the Drakensberg, and so on to the Free State uplands. A similar trip by the Delagoa Bay or Beira Railway would show the

same rising of the coast belt into a moderately-elevated terrace, which again forms the pedestal for a still higher terrace, that in turn merges into the vast central plateau. An orographical map of South Africa shows these successive plateaux very clearly. First, there is what may be termed the *Coast Plateau*, a belt of land rising from the seaboard to an average height of about 600 feet, and varying in width from a few miles to fifty miles or so. This plateau adjoins the west and south coasts, and may be said to end at Cape St. Francis. Further east, long swelling uplands and forest-clad mountains come down almost to the water's edge, and the bold and rocky coast rises here and there into huge cliffs, the most notable being those which tower on either side of the entrance to the St. John's River.

North of Natal, the coast belt gradually broadens out and finally merges into the low-lying plains of the Lower Zambesi. Besides the coast towns already mentioned, the following inland towns are in this district :—*Malmesbury, Wellington, Paarl, Stellenbosch, Caledon, Swellendam, Riversdale, Humansdorp,* and *Uitenhage.**

The Coast Plateau is divided from the next terrace— the plateau of the *Southern Karroo*—by the Lange Bergen and other ranges ; and this again from the third plateau—the great upland of the central or *Great Karroo* —by the Zwartebergen or Black Mountains. The towns in the Southern Karroo district are *Worcester, Montagu, Robertson, Ladismith, Oudtshoorn* and *Uniondale.* The Great Karroo extends east and west for about 350 miles at a level of from 2,500 to 3,500 feet above the sea. Along the northern border of this great upland basin rises the long and comparatively lofty range which, under various names, stretches from the hills of Little Namaqualand to the lofty range of the Drakensberg. The central range, the Nieuwveld, is flanked by the Roggeveld on the west, and by the Winterberge, the Sneeuwberge, and the Stormberge on the east—the culminating point of this bold escarpment of flat-topped heights being the Compass Berg, 7,800 feet, in the Sneeuwberge range. This long range forms the central watershed or waterparting of the Cape —the drainage on the one side flowing north to the Orange River, and on the other, south into the Indian Ocean. The chief towns in this section are *Beaufort West, Prince Albert, Willowmore, Graaf Reinet, Somerset East* and *Aberdeen.*

Beyond this great range stretch the vast and scantily-watered uplands of the *Northern Karroo,* the loftiest and most extensive of all the plateaux to the south of the Orange River. This plateau has an average elevation of about 3,000 feet, and upon it are situate the towns of *Cradock, Queenstown,* and *Tarkastad ; Aliwal North, Burghersdorp, Dordrecht ; Colesberg, Richmond, Victoria West, Carnarvon,* and *Fraserburg.*

Under the general name of *Eastern Uplands* may be included all the broken middle terrace country from Graham's Town to the Manica upland. In this district are *Graham's Town, Fort Beaufort, Stutterheim, King William's Town* and *Bedford,* in the Cape Colony ; *Umtata,*

in the Transkei ; *Pieter-Maritzburg, Howick, Estcourt* and *Ladysmith,* in Natal ; *Vryheid,* in Dutch Zululand ; and *Bremersdorp,* in Swaziland.

These Eastern Uplands are divided by the long range of the Drakensberg and its continuations from the great *Eastern Plateau,* which includes the mountainous Basuto country, the higher upland plains of the Free State, and the hilly and undulating "Hooge Veldt" of the Transvaal. The Drakensberg or Quathlamba* Mountains form the "backbone" of South Africa ; and although the loftiest summits do not reach the line of perpetual snow, the range attains—in Giant's Castle—an elevation of nearly 10,000 feet, and still higher in Champagne Castle and Mont aux Sources, 11,000 feet. The highlands along the edge of the Mashona plateau rise in Weelza to 5,400 feet, and in Sadza to 4,500 feet—two peaks at the sources of the Sabi River. The plateau itself is about 4,000 feet above the sea. Johannesburg, which stands on the highest ridge of the High Veldt of the Transvaal, is at an elevation of 5,600 feet.

The Northern Karroo of the Cape Colony, the great Eastern Plateau of the Free State and the Transvaal, with its continuation, the Matabele and Mashona plateaux, on the one side, and, on the other, the broken uplands of German South-West Africa, all slope gradually towards the vast plains of Bechuanaland and the Kalahari Desert. From the Orange River, northwards to about the 22nd degree of latitude, Bechuanaland is mainly a broken plateau of 4,000 to 5,000 feet in height above the sea, dry and devoid of perennial streams,† but beyond that it slopes gradually down to the basin of Lake Ngami, which is little more than 2,000 feet above the sea.

The want of rain, or rather the spasmodic and violent character of the rainfall, being the chief drawback of the greater part of South Africa, the *rivers* are of necessity an unsatisfactory feature in the physical conditions of the country. Many of them are periodical streams, flooded to excess after the rains, speedily drying up, and becoming mere chains of pools in the dry season. There are, of course, numerous perennial rivers, but all of them are similarly liable to great and sudden variations in volume. In an hour or two after a heavy thunderstorm, the most insignificant stream becomes a raging torrent of turbid water, rushing impetuously between its steep banks, perhaps overflowing them and flooding the adjoining veldt. But the flood is as brief as it is violent, and it is very rarely that the transport-rider is forced to outspan on the banks of a swollen stream for more than a few hours, or a few days at the longest.

The rivers improve, and the flow of water increases, as we proceed eastwards. Instead of the dry water-courses and sand-rivers that furrow the solitudes of the western coastlands, we are charmed with the delightful babble of never-failing streams, that run down the verdant mountain slopes and wind through many a wooded kloof. Generally speaking, however, the volume of water in

---

* Full details of these and all other South African towns are given in the *Castle Line Guide to South Africa.*

* A Kaffir name, meaning "heaped up in a jagged manner."

† See further, Chap. IX., *Livingstone and Central Africa.* By H. H. Johnston, C.B., H.M. Commissioner for British Central Africa. The World's Explorers Series. (London : G. Philip & Son).

South African rivers is comparatively small; and when, swollen by rains, they do attain a respectable size, the flood water runs off very quickly, and the river soon shrinks into its ordinarily insignificant dimensions.

Besides their normally-limited volume of water and liability to sudden and dangerous floods, the courses of all South African rivers are, owing to the peculiar terrace-like conformation of the country, repeatedly broken by waterfalls or rapids, which, of course, mark the more or less abrupt stages in their descent from terrace to terrace. The streams also, especially when in flood, bring down such enormous quantities of sediment that, in nearly all cases, "bars" have been formed at their mouths, which prevent, or greatly impede, the entrance of sea-going vessels. South African rivers are therefore, on the whole, of but little value as waterways into the interior.

The "bars" at the river-mouths may, however, by the aid of breakwaters and training walls, be to some extent removed, and the estuaries thus converted into safe and convenient harbours. This has been done with considerable success at East London, where vessels of 4,000 tons can now anchor in the river. About half a million sterling has been spent on similar work at the Kowie, but not much improvement has been effected. In Natal, also, the energy and enterprise of the Harbour Board have so far overcome the peculiar difficulties caused by the ebb working on the sandy bottom of that part of the coast, that the "bar" at the entrance to the port is scarcely ever impassable, and vessels of considerable size, such as the s.s. *Dunrobin Castle*, can now enter the bay. The Orange and the Limpopo are similarly blocked at their mouths, although the Orange, like several of the south and east coast rivers, is navigable for small craft for some miles inside the bar. The Zambesi, which has, as it were, been politically annexed to South Africa, is navigable for gunboats and river-steamers for about 260 miles from the sea. Beyond that, numerous rapids and falls, especially the stupendous Victoria Falls, render this magnificent river useless as a waterway into the far interior. Major Serpa Pinto says that, between the 16th parallel and the Falls, its channel is obstructed by no less than 72 cataracts and rapids. The navigability of the lower Zambesi derives additional importance from the fact that its great tributary, the Shiré river, which flows from Lake Nyasa, is also navigable, except at one point —the Murchison Falls.

In past times, South Africa was, no doubt, a country of great lakes and vast inland seas. Dr. Livingstone's theory that the karroos and other large plains once formed the beds of immense lakes is strongly corroborated by the fossil remains found there. The rims of these great basins, it is supposed, were fissured or cracked by up-heaval at a comparatively recent geological period, and through these fissures the waters were discharged. "The fissures thus made at the Victoria Falls let out the waters of the great Zambesi Lake. The fissure through which the Orange River pours itself at the Falls of Aughrabies probably drained off the waters that then covered the Kalahari and the table-lands of Bushmanland. The Warm Bokkeveldt valley and Kannaland, as well as the Great Karroo itself, were evidently lakes at one period,

their waters escaping by the fissures of Mitchell's Pass, the Gauritz, and the Hex River Valley; and, indeed, the rugged and fearful kloofs, through which their surface waters still escape, show the evident traces of some violent convulsion of nature. The basins of Cradock and Queens-town, evidently old lake-beds, are now drained by the water-courses of the Great Fish and Kei rivers."*

This theory finds perhaps its strongest corroboration in the fact that the remnant of the great Kalahari basin —the shallow Lake Ngami—is gradually drying up. When discovered by Dr. Livingstone in 1846, it was about 50 miles in length and eight or ten miles in width. It receives the Cubango from the inner uplands of Angola, and in the rainy season this stream pours such a flood of water into the shallow basin of the Ngami that it over-flows by the Botletle, or Zouga channel, into the great vleis or saltpans on the east. These "vleis" are shallow sheets of water, which, after heavy rains, accumulate in natural hollows in the ground, and on evaporation leave an incrustation of salt on the surface, and are hence called saltpans. Andersson's Vlei, in the Kalahari, to the east of Lake Ngami, the Commissioner's Saltpan in Great Bushmanland, the four vleis which form the sheet called the Groen Vlei on the coast in the Knysna district, are the largest of these variable lakes. There are hundreds of "fonteins," or fountains, of delicious water distributed over the country, and numerous hot or mineral springs, some of which have medicinal properties of high value.

Physically, then, South Africa is a distinct and homogeneous region, solid and unbroken; but, politically, the country is split up into a conglomeration of states and territories, colonies and protectorates, under a dozen different kinds of government and forms of administration —from the most primitive to the most advanced, and from the most democratic to the most absolute and despotic. We have British Colonies, with full responsible government; British Crown Colonies and Protectorates, under the direct control of the mother-country; a great British Company, with extraordinary powers over vast areas; two independent Boer Republics; an enormous German Protectorate; and an immense Portuguese Dependency: to say nothing of native chiefs innumerable, who, nominally subject to British, Boer, Teutonic, or Lusitanian control, still rule their tribes and clans in patriarchal fashion—a veritable political mosaic, curious, but rather puzzling. The following notes, with a careful examination of the map, will, it is hoped, give the key to the puzzle, and enable the reader to gain a clear idea of the actual and relative position and extent of the various political divisions of Southern Africa.

Until the recent expansion of British authority northwards to and beyond the Zambesi, the Cape Colony was the most extensive, as it still is the wealthiest and most important, of all the territorial divisions of Southern Africa. The colony takes its name from that of the famous headland discovered by the disappointed Portuguese navigator, Bartholomew Diaz, in 1486, and by him named "El Cabo de todos tormentos"— "the Cape of all the

---

* Silver's *Handbook to South Africa*.

Storms "—in remembrance of the exceptionally severe weather which he encountered off the coast, while making his futile attempt to reach India by sea. The Portuguese monarch, on hearing Diaz's report, and rightly judging that the discovery gave "good hope" of ultimate success, changed the name which the baffled navigator had given to it, into the very opposite one of " El Cabo de Boā Esperanza," the *Cape of Good Hope*—a name which the headland has ever since borne, and which is still the official name of the colony.

Cape Colony occupies the southern extremity of the continent. Its western shores are washed by the waters of the Atlantic, and its southern and south-eastern coasts by those of the Indian Ocean ; the nominal boundary between the two great oceans being the 20th meridian of East longitude—the meridian of Cape Agulhas, the southernmost point of the Cape and of Africa. The colony extends northwards to, and, since the incorporation of *Griqualand West*, beyond the Orange River ; eastwards, by the gradual absorption of the Kaffrarian territories (a process which was completed by the recent annexation of *Pondoland*), the Cape is conterminous with Natal, the boundary being the river Umtamfuna. Of the formerly quasi-independent territories in Kaffraria, *Pondoland* and *Griqualand East* adjoin Natal, while *Tembuland* and the *Transkeian Territories*—Fingoland, the Idutywa Reserve, and Gcalekaland—extend along the Kei River, the former boundary of the Cape Colony. Twelve islands off the coast of Damaraland, together with the port of Walfish Bay—the only good harbour on the entire seaboard of German South-West Africa— also belong to the Cape. On the north-east of the colony lies *Basutoland*, now a separate colony, but from 1871 to 1884 a part of the Cape Colony. To the north are the two Dutch Republics, Bechuanaland, and the German Protectorate of South-West Africa. How large a territory is included in the colony may be inferred from the fact that a bee-line of 600 miles may be drawn across it from east to west, and one of 450 miles from north to south, and that its area of over 230,000 square miles is four times that of England and Wales, or twice that of the whole of the United Kingdom.

Natal lies to the north-east of the Cape Colony, and includes the pleasant and verdant land between the Umtamfuna and the Tugela, discovered on Christmas Day, 1497, by Vasco da Gama, on his celebrated voyage to India, and therefore named by him *Terra Natalis*. Natal has an area of about 20,000 square miles—about that of England and Wales, and one-twelfth that of the Cape—and extends inland to the giant range of the Drakensberg, which rises from 80 to 120 miles distant from the coast, and separates Natal from Basutoland and the Orange Free State.

Basutoland, the "Switzerland of South Africa," is a mountainous country, 10,000 square miles in extent— about twice the size of Yorkshire—and completely hemmed in by the Cape, Natal, and the Orange Free State. Since 1884, when it was disannexed from the Cape, it has formed a British Crown Colony.

Another British Crown Colony is *Zululand*, or rather the remnant of Zululand saved in 1887 from absorption along with the rest of the country into the Transvaal. To the north of Zululand are two other native territories —*Tongaland*, a British protectorate, on the coast, and *Swaziland*, an inland state, still under the joint protection of Great Britain and the Transvaal, but in all probability soon to be annexed to the latter.

The twin Dutch republics are also entirely inland—the more southerly, the *Orange Free State*, being divided from Cape Colony partly by the Orange River and partly by an artificial boundary which passes close by the " Diamond Fields " to the Vaal River, which separates it from the *South African Republic* or the Transvaal. The Free State is about four-fifths the size of England and Wales, but the Transvaal is twice as large, extending from the Vaal on the south to the Limpopo or Crocodile River on the north.

To the west of the South African Republic extend the great plains of the British Colony and Protectorate of *Bechuanaland*, with an area of 170,000 square miles, or three times that of England and Wales. North of Bechuanaland and the Transvaal, and extending to and beyond the Zambesi, are the enormous territories, the administration and development of which have been entrusted, and so far with signal success, to the *British South Africa Company*, of which Mr. Rhodes is the moving spirit. The operations of the company now cover the whole of inner South-Central Africa from Mafeking to Tanganyika—a territory three-quarters of a million square miles in area, or nearly nine times the size of Great Britain. The Zambesi divides this region into two great sections, which may be distinguished as *Northern Zambesia*, or *British Central Africa*, and *Southern Zambesia*. Northern Zambesia includes the *Nyasaland Protectorate*, the Barotse country, and other undeveloped and but imperfectly-known territories. Southern Zambesia includes *Mashonaland* and *Matabeleland*, both of which countries now loom large in the public eye, although, but a year or two since, they were practically unknown and inaccessible, except to a few adventurous travellers or daring traders.

There are yet two other territories in Southern Africa that must be noted, both extensive, and both under foreign domination. To the German Protectorate of *South-West Africa* belongs the entire region between the Orange River and the Cunene, with the exception of a small territory on and around Walfish Bay, and a few islands off the coast which belong to the Cape. The German territory, which has an area of not less than 350,000 square miles, or one-third larger than the German Empire in Europe, extends inland to the frontier of Bechuanaland, with, on the extreme north-east, a narrow strip extending along the Chobe Valley to the Zambesi.

On the opposite side of the continent another foreign power—Portugal—exercises a feeble authority over immense territories along the coast to the north and south of the Zambesi, and over a limited belt of country on either side of the river as far as Zumbo, about 600 miles from the coast—an area altogether of over a quarter of a million square miles, which, however, is only partly within the limits of Southern Africa.

III.—A LAND OF SUNSHINE AND HEALTH.

Three Africas known to the modern world—Sunny South Africa—The light, rich and brilliant—The air, clear and transparent—A fine climate—Heat not excessive—Large daily range of temperature—Temperate South Africa—The gradual rise of the country from the coast, and corresponding reduction of temperature—The temperature of the various divisions of the country—The rainfall—"It never rains but it pours"—Reckless destruction of the bush—Irrigation in its infancy—Droughts—Olive Schreiner's graphic picture—Annie Martin's description of a drought in the Karroo—Rainfall in the Cape, Natal, the Republics, etc.—Health Resorts—The voyage to the Cape.

There is a common saying to the effect that three distinct Africas are known to the modern world—North Africa, where men go for health ; South Africa, where they go for wealth ; and Central Africa, where they go for adventure. The statement is certainly as true as it is clever ; but, as far as South Africa is concerned, it does not express the whole truth, for men not only go to the "Land of Diamonds and Gold" for *wealth*, but they go in ever-increasing numbers to the "Land of Sunshine" and balmy air for *health ;* and, as we shall presently see, men go there for *adventure* also, and find in the interior and towards the Zambesi a veritable "Sportsman's Paradise," and enough excitement, spiced with danger, to tingle the nerves of the most hardened hunter. Health, wealth, excitement—the world's three-fold desire—may be found in "Sunny South Africa."

"Sunny South Africa" is, in truth, a Land of Sunshine. The light is as rich and brilliant as the air is surpassingly bright and clear—a striking contrast to the diffused light and hazy atmosphere of cloudy England. This difference has almost a startling effect on the newcomer, particularly if he happens to enter Table Bay on a clear Cape winter morning. Looking landward, he sees "the long range of mountains, which completely separate the peninsula from the mainland, though at a distance of from seventy to a hundred miles, standing out with a sharply-defined outline—the ravines, and watercourses, and terraced heights appearing with almost supernatural distinctness." A stranger, describing his first impressions on landing in the country,* says that "the characteristic beauty of light in South Africa is not seen in its marriage with manifold forms of cloud, so much as in the full and even splendour with which it penetrates that. Distant objects that, in a less brilliant atmosphere, fade away in hazy outline, stand out with perfect distinctness. Small boulders, cavernous hollows in the rocks, patches of bush at the head of the kloofs, at an elevation of two or three thousand feet, are seen without difficulty. Let the spectator place himself a distance of twenty or thirty miles from Table Mountain or the Katberg in South Africa, and then do the same with Snowdon or Cader Idris in the mother country, and he will be surprised at the contrast in the aërial perspective. The two latter elevations will appear in more or less of hazy outline, with details of face and profile obscured ; but in the clear atmosphere of South Africa, the direction of the watercourses, the curves of the kloofs, and indeed every bold wrinkle on the face or slope of the mountains, will be clearly discerned. I have sometimes," he adds, "looked at Table Mountain at what photographers would call the sharp definition of every line, until the sense of distance almost vanished, and it has seemed as if I must see a human figure if it were climbing the heights,

* *Official Handbook to the Cape and South Africa.*

or hear a human voice if it broke the silence of the kloofs." *

South Africa has, undoubtedly, one of the finest climates in the world ; but in so extensive a country, with such diversity of altitude and form, we necessarily find considerable differences in the climate in different parts of it ; on the whole, however, it may be said to be temperate, dry, and healthy. The country is, of course, much warmer than Great Britain, but the heat is nowhere excessive, and though the direct rays of the sun may be extremely powerful, particularly in summer, yet the peculiar dryness and rarefaction of the air make even the occasionally intense heat easily bearable. "In some of the deep-lying valleys, where the motionless air becomes heated by the large mountain masses, the heat is, in summer, oppressive, but the actual heat is at no time excessive." Dr. Lawrence Herman, in his paper on "The Cape as a Health Resort,"† gives, as an example, Kimberley, notably one of the hottest places in the country, with a maximum temperature ranging from about 75° F. in June, to 105° F. in January ; "and yet," he says, "there is no place in the Cape where people have more ceaseless activity, or more restless energy. Europeans work all day, heedless of the heat. The day is characterised by a maximum of sunlight, a balmy, buoyant atmosphere, with a clear, cloudless sky of the purest blue, and a cool night succeeds a warm day."

The seasons in South Africa are, of course, the reverse of those in Europe, but they are not so well marked, and it is only in some parts of the coast that the difference between spring and summer, autumn and winter, can be traced. The country generally may be said to have only two seasons—the warm and the cool, or summer and winter. In summer, which in South Africa may be said to extend from October to March, it is considerably hotter than in England at the corresponding season, but summer in England is often much more oppressive than it is in South Africa, on account of the moisture which the air of our beclouded land contains. In the coldest South African winter weather, even in the up-country districts, the sun is hot ; in fact, winter, in the English meaning of the term, is unknown. Compared with ordinary English temperatures, South Africa is certainly a warm country, but scarcely anywhere is there anything approaching the excessive heat and humidity that makes the climate of India so debilitating to Europeans. Formerly, indeed, before the Suez Canal was opened, the Cape was the favourite resort of invalids from India, and no climate in the world could be more agreeable to the feelings, and very few more beneficial for the usual class of Indian invalids, than a Cape winter. There is an in-

* Quoted in Silver's *Handbook to South Africa*, p. 665.
† *Official Handbook to the Cape and South Africa.*

vigorating freshness about this season equally delightful and beneficial ; the moment the rain ceases, the clouds rapidly clear away, and the sky remains bright for several days.*

Another striking feature in the climate of South Africa is the large daily range of temperature. But the result of these apparently trying and sudden changes of temperature is, if ordinary precautions are taken, most beneficial, inasmuch as the heat of the day is never prolonged into the night, and so causing exhaustion and preventing sleep. All over South Africa, summer and winter alike, the nights are cool and refreshing, and on the uplands and mountain districts, decidedly cold, and not infrequently frosty. Snow generally falls on the mountains, and sometimes on the higher plains, but nowhere does it remain all the year round, the loftiest mountains being far below the snow-line.

The climate of South Africa is, on the whole, far more temperate than that of countries within the corresponding parallels north of the equator. Cape Town, for instance, is only about 34 degrees distant from the equator, and yet it has a mean annual temperature of 62° F., which is about the same as that of Naples, Nice, and the Riviera, in from 41 to 43 degrees north of the line. This is due to the fact that "Temperate South Africa" is surrounded by vast, open oceans, and is swept by cool winds from the cold regions of the Antarctic Seas, from which also a cold drift impinges against the western coast.

The climate of South Africa is also powerfully affected by the peculiar conformation of the country, by which its entire surface is exposed, step by step, to the cooling sea breezes and strong gales which sweep freely over it ; and by the ever-increasing altitude of its terraces and plains —the land rising rapidly, in many places from the very margin of the sea, to the great ranges which are—as Mr. Russell, in his excellent book on Natal, points out—the rugged cliff-edge of the crowning terrace, the vast central plateau of South Africa. Taking Natal as an example of the step-by-step structure of South Africa, we see that, "from the sea to the Drakensberg, the land rises by successive terraces, well-known to travellers on the main road between the Port and Van Reenen's Pass. The first terrace, 1,730 feet high, rises above the village of Pinetown, 12 miles inland ; the second, 2,424 feet high, is at Botha's Hill ; the third, 3,700 feet high, begins on the town hill, above Maritzburg, 45 miles from the sea ; and the fourth, 5,000 feet high, forms the highlands between the villages of Weston and Estcourt. From this point, the surface rises and falls with little variation till the Pass is reached, and at a distance of 225 miles from Durban by rail." †

In Cape Colony, the land rises similarly in successive steps from the seaboard to the Orange River—the first step, or Coast Plateau, averaging about 600 feet in height ; the second terrace, the Southern Karroo, and the Warm Bokkeveldt, rising to between 1,000 and 2,000 feet ; the

third plateau, being the Central or Great Karroo, with an average altitude of about 3,000 feet ; and the fourth, the still loftier Northern Karroo, the highest and most extensive of the Cape plateaux, varying in height from about 2,700 to 6,000 feet above the sea-level, the average elevation being between 3,000 and 4,000 feet. Beyond the Orange stretches the Diamond Fields country, a less elevated continuation of the Northern Free State ; the wide undulating plains of the Free State, with an altitude of from 4,000 to 5,000 feet, and over 50,000 square miles in extent ; the loftier and still more extensive plateaux of the Transvaal, with a general elevation of from 5,000 to 7,000 feet ; the lower but more diversified uplands of the Matabele country ; and, on the opposite side of the continent, the irregular highlands of Damaraland and Namaqualand ; these, together with the Cunene Tableland and the trans-Zambesian Highlands on the north, encompass and merge indefinitely into the great central plains of Bechuanaland and the dreary expanse of the Kalahari Desert. There are, as we have already remarked, indications that this great central basin formed the bed, at some remote age, of an immense inland sea, whose waters rushed in intermittent mighty cataracts over its mountain-rim, as its rocky floor was gradually raised by a succession of volcanic disturbances. Its present elevation is about 4,000 feet. Vryburg, in the south, is 4,300 feet, and Lake Ngami, in the north, 3,700 feet above the level of the sea. When we bear in mind that the thermometer falls one degree for every 300 feet of altitude, we can readily see how, even in the semi-tropical zone of the country, the climate is comparatively cool and bracing ; and how, in the winter, on the high tablelands of the Karroos, the Free State, and the Transvaal, the nights are often intensely cold, the temperature frequently falling to some degrees below freezing point.

Throughout South Africa, the temperature is more equable in the coast districts than on the uplands in the interior ; the mean daily range for the year between maximum and minimum being about 15° on the coast, but nearly double that amount in the Free State and at the Diamond Fields. Again, while the mean annual temperature of the Diamond Fields and the coast, as far as Natal, is about the same (63° F.), the mean maximum for the month, which reaches 80° F. on the coast, frequently exceeds 90° F. at Kimberley. North of Natal, the coastlands become much hotter, and Durban itself has a mean annual temperature of 8° F. above that of Cape Town, where the mean temperature for the year is almost exactly the same as the mean summer temperature of England, 62° F.

On the whole, then, we may say that the coast climate of the Cape Colony is warm, moist, and equable ; that of the midland is colder and drier in winter and hotter in summer ; the mountain climate is drier still and more bracing, but with extremes of heat by day and cold by night. The hottest month is generally January ; during that month the average maximum temperature during the last ten or twelve years at Cape Town was 82° F., at Port Elizabeth 76° F., at Graaf Reinet 87° F., at King William's Town 84° F., Aliwal North 84° F., Clanwilliam 93° F., and Kimberley 93.7° F., which is the highest

---

* Dr. Stovell, in the *Bombay Medical Journal.*

† See further, *Natal : The Land and its Story,* by Robert Russell, Superintendent of Education. (Maritzburg : Davis and Sons. London : Simpkin, Marshall and Co.).

B

average in the colony. July is generally the coldest month, and the minimum average at Cape Town is 46° F., at Port Elizabeth 48° F., at Graaf Reinet 36° F., at King William's Town 34° F., Aliwal North 28° F., Clanwilliam 39° F., and at Kimberley 38° F. The fierceness of the sun's heat must be taken into account in judging of the temperature of the country, but the dryness of the air makes the heat less felt than it would be in a damper climate.*

The climate of Natal varies considerably, but although it is nearer to the tropics, its mean annual temperature hardly exceeds that of the Cape, which is nearly the same as the mean summer temperature of England. At Durban the highest temperature in the shade, in 1892, was 105° F., and on 31 days of that year it was over 90° F.; the lowest was 47° F. in August, and on 9 days only was it below 50° F., while in January, February, March, and December it was never below 60° F. The mean temperature at Pietermaritzburg is on an average of years between 64° and 65°, or 4° lower than on the coast. The winter is bright and dry, and the summer heat is tempered by a clouded sky and frequent thunderstorms and heavy rains. The climate is much more bracing at places like Newcastle, Dundee, Howick, Pinetown, and Estcourt, than it is on the coast; the summers are hot, but the nights and winters are cold. Snow is very rarely seen, except on the tops of the highest hills and mountains, but in the uplands the temperature on some nights in the winter falls below the freezing point.†

In Bechuanaland, owing to the elevation, the extremes of heat by day and cold by night are great, especially in the winter, which, according to Dr. Livingstone, who spent many years in the country, is the complete antipodes of our cold, damp English winter. Here "the winter is perfectly dry, and as not a drop of rain even falls from the end of May to the beginning of August, damp and cold are never combined. However hot the day may have been at Kolobeng—and the thermometer, previous to rain, sometimes rose to 96° F. in the shade—yet the atmosphere never had that straining and debilitating effect so well known in India and in parts of the coast regions of South Africa itself. You may sleep out of doors with the most perfect impunity, as for many months not a drop of dew falls."

The climate of the Free State, says Dr. Lawrence Herman of Cape Town,‡ is most delightful, being cool and bracing, with a bright, superabundant, almost dazzling sunlight. The days in summer during the morning and evening are cool, but in the middle of the day the heat is considerable, while the nights are cool and refreshing. During the winter, the air is balmy and warm in the sun, but the nights are intensely cold, the temperature rapidly falling to some degrees below freezing point. Dr. Fuller, of Kimberley, writing of the same country, says that,

during the six hottest months of the year, the average maximum temperature is 82° F., the average minimum for the same period 55° F., and the highest for one month being 60° F. The heat of summer, therefore, he adds, is considerable, but perfectly tolerable with the dry atmosphere; the nights are deliciously balmy, and enable an invalid to sleep with doors and windows open during the night, or even to sleep altogether in the open air.

The Transvaal, although partly within the tropics, is so considerably elevated—the general elevation of the country being from about 5,000 to 7,000 feet—that the climate is cool and bracing, much more so than the coast-climate many degrees further south. The winter season, from April to September, is, says Mr. Jeppe, cold and dry, particularly during the nights; the days are often as warm as in summer. During the winter months, cutting, sharp, cold winds blow from the south, and the High Veldt and the Drakensberg Mountains are frequently covered with snow. The mean annual temperature is 68·64° F., or about 6½° higher than that of the Cape, or the mean summer temperature of England. At Johannesburg, Mr. Miles, C.E., found the mean temperature for the three summer months—December, January, and February—was, in 1889, 72·91° F., while for the three winter months—June, July, and August—it was 52·74° F. "During the winter months great and sudden changes are experienced, for example, in May, 1892, on two days the temperature in the sun fell from 111° F. to 30° F.; in June, on 19 days from over 100° F. to under 4° F., and on 14 days from over 100° F. to freezing point; in July, on 18 days from over 100° F. to freezing point. In August, on twelve days from over 100° F. to under 40° F., and on five days from over 100° F. to freezing."*

In the mountain districts of the Cape and Basutoland, the climate is very fine, exhilarating in summer and intensely cold and most bracing in winter. Snow lies on the Cold Bokkeveldt in the Cape for weeks at a time, as it also does on the Stormberg and other mountains. In the Basuto country, the days in summer are warm, but the heat is never excessive; while in winter even the days are cold, and the nights intensely so. "Frost is generally met with towards the end of April; May is a delightful month; and June is the commencement of winter."

Although entirely within the tropics, the climate of Mashonaland and Matabeleland is by no means tropical, the temperature ranging from 36° to 86° F.; while beyond the Zambesi, on the Milanji plateau, the heat, though very considerable, is not excessive or injurious to Europeans, provided ordinary precautions are taken, and their stay is not too prolonged.

In German South-West Africa, the climate of the Damara uplands in the interior, which have an average elevation of from 3,000 to 5,000 feet, is not too warm, and is even cold in winter. But along the arid and waterless coastlands from the Cunene to the Orange the climate is not agreeable to Europeans; the heat is very great, but it is dry, and a much more intense degree of

---

* Handbook to the Cape, issued by the Emigrants' Information Office.

† Handbook to Natal, issued by the Emigrants' Information Office.

‡ In his Paper on "South Africa—its Climate and Health Resorts," in Mr. John Noble's Official Handbook to the Cape and South Africa.

* Noble's Official Handbook.

dry heat can be borne without injury than the moist and enervating heat that characterises the coastlands on the opposite side of the continent.

Dryness is the chief characteristic of the climate of South Africa, but, when it does rain, "it pours." Vast torrents of rain fall in a very short space of time, as much falling in an hour as in a day in England. This, as Dr. Fuller of Kimberley points out, exerts a marked influence on the humidity of the atmosphere, which is much lessened for the whole day with the more rapid downpour. And not only is the rainfall intermittent and violent in character, but it varies considerably in amount in different parts of the country and from year to year, while periodical and long continued droughts greatly interfere with agricultural operations even in districts where irrigation is possible. Except on the Eastern coastlands and in the neighbourhood of the Knysna and other forest tracts, the surface of the country generally is so hard that the torrential rains flow off as fast as they fall, there being nothing to restrain the moisture and allow of slow filtration.

The statement is frequently made that South Africa is drying up. If by this, says an eminent hydraulic engineer,* is meant that the springs and streams are not so constant as they used to be, the statement is undoubtedly true. If it is meant that less rain falls now than in former historic times, the statement has certainly not been proved, and is most probably untrue. The early books of travel speak of droughts in the interior; Sparrman mentions the great drought of 1775. Careful rainfall measurement at the Royal Observatory at Cape Town gives no support to the view that the rainfall is diminishing. But the cutting down of trees and the burning of the veldt have affected, and are affecting, the permanence of springs and streams. Both white men and natives seem to act recklessly in this matter, cutting down bush for kraals and firewood, the natives, especially, using large quantities of young trees for their tents and game traps. The increased number of flocks has also contributed to this result. Where the grasses and bushes are eaten off, the sun bakes the soil, and the rain runs off into the rivers, forming new "sluits" as it runs, and is lost in the sea without replenishing the underground supplies. Sir Charles Warren thinks that the alleged decrease of rainfall may be due to a gradual change of grasses on the veldt, and to the introduction of sheep, as well as to the veldt fires. In former days, he says, there were long grasses which were not suited for sheep, the sun scarcely ever reached the soil, and evaporation was, therefore, very gradual; consequently the soil remained damp, and there were many vleis and pans of water. Since the introduction of sheep, the sun has been able to beat fiercely upon the soil, moisture is rapidly exhausted, and, in addition to this, the numerous cattle-tracks tend to carry off the rainfall much more rapidly to the sluits and rivers, and the river beds have, in many cases, sunk many feet during the last fifty years. The result is that, where there used to be morasses and swamps, there are now dry watercourses, and this is very remark-

able in many parts of the country.* During his stay in South Africa, Sir Charles Warren also frequently noticed that, where rain fell on a piece of ground early in the season, succeeding showers fell on the same piece of ground, while adjoining farms remained comparatively dry; and it frequently occurred that, where a large tract became wet, heavy rains continued to fall during the season. Of this peculiarity of the rainfall in South Africa, Mrs. Annie Martin, the accomplished authoress of "Home Life on an Ostrich Farm"†—a series of most delightfully piquant and graphic sketches of life in the Karroo—says :—"The partiality of the thunderstorms is surprising; sometimes one farm will have all its dams filled, while another near it does not get a drop of rain. Often, during a whole season, the thunderclouds will follow the same course, one unlucky place being repeatedly left out."

In almost all parts of "dry" South Africa, there are immense tracts of the most fertile land, which only require water to produce the most abundant crops. Numerous wells have been sunk and dams made, but irrigation is as yet in its infancy, though destined, in the near future, to change the face and the fortunes of the country, and to enable it to support with ease a population from ten to twenty times as numerous as it does at present. In the meantime, "the long droughts are certainly very trying; indeed, they could not possibly be endured by any country less wonderfully fertile than South Africa, where it is calculated that three good days' rain in the year, could this but be had regularly, would be sufficient to meet all the needs of the land. But often, for more than a year, there will be no rain worth mentioning; the dams, or large artificial reservoirs, of which each farm usually possesses several, gradually become dry, and the veldt daily loses more of its verdure, till at last all is one dull, ugly brown, and the whole plain lies parched and burnt up under a sky, from which every atom of moisture seems to have departed—a hard, grey, metallic sky, as different as possible from the rich, deep-blue canopy, which, far away to the north, spreads over lovely Algeria. The stock, with the pathetic tameness of thirst, come from all parts of the farm to congregate round the house, the inquiring ostriches tapping with their bills on the windows as they look in at you, and the cattle lowing in piteous appeal for water; and you realise very vividly the force of such scriptural expressions as, 'the heaven was shut up,' or, 'a dry and thirsty land, where no water is.'"

Olive Schreiner, in her incomparable "Story of an African Farm," thus describes the great drought of 1862. "From end to end of the land, the earth cried for water. Man and beast turned their eyes to the pitiless sky, that, like the roof of some brazen oven, arched overhead. On the farm, day after day, month after month, the water in the dams fell lower and lower; the sheep died in the fields; the cattle, scarcely able to crawl, tottered as they moved from spot to spot in search of food. Week after week, month after month, the sun looked down from the

---

* *J. G. Gamble, Esq., C.E.*

* *Proceedings of the Royal Colonial Institute,* vol. xvii., p. 9.

† Published by Messrs. George Philip & Son. London and Liverpool.

cloudless sky, till the Karroo-bushes were leafless sticks broken into the earth, and the earth itself was naked and bare, and only the milk-bushes, like old hags, pointed their shrivelled fingers heavenward, praying for the rain that never came."

The eastern coastlands are much more favourably situated with regard to water than the rest of the country. Their mountain-slopes and valleys are clad in verdure, kept ever green by the moisture carried to them by the trade-winds of the Indian Ocean. The vapour-laden clouds and humid air-currents are arrested by the encircling mountain-barriers, and the moisture thus condensed descends in refreshing and fertilising showers on their seaward sides. The Cape Peninsula, and some parts of the south coast, together with many favoured spots in the interior and towards the Zambesi, have an abundant rainfall, and, consequently, a luxuriant vegetation.

In the eastern portion of the country, the wet season is the summer; in the western, the winter. In the summer, the south-easterly winds, laden with the moisture from the warm waters of the Indian Ocean, are caught by the high mountain-chains and plateaux and deprived of their moisture; so that, while at this season heavy rains fall in the eastern division of the Cape, the Transvaal, and Natal, the western districts are comparatively dry. In the winter, on the other hand, the prevailing north-westerly winds discharge their moisture in copious rains over the western districts, the midland and eastern districts being then usually dry.

Mr. Gamble thus sums up the results of the meteorological observations at the Cape. The north-west of the colony, he says, is almost rainless. The south-west has abundant winter rains. The south coast has rain in all months, December and January being the driest time; in the midlands, as well as in the north and east, the rains occur generally in February and March, although near the coast there is a second maximum in October and November. Droughts seldom occur all over the colony in the same year; in fact, it seems as if a drought in the interior frequently occurs in the same year as abundant rains on the south-west coast.

In the Cape Colony, the annual rainfall (in inches) during the last nine years has ranged from 23 to 41 near Cape Town, 8 to 15 at Worcester, 16 to 24 at Mossel Bay, 6 to 13 at Oudtshoorn, 12 to 24 at Graaf Reinet, 14 to 27 at Colesberg, 12 to 30 at Kimberley, 26 to 42 at Graham's Town, 17 to 36 at King William's Town, 16 to 28 at Port Elizabeth, 15 to 41 at Queenstown, and 2 to 12 at Beaufort West, as against 26 inches in London. In the Great Karroo and Great Namaqualand and other districts in the north-west, less than 6 inches of rain fall in the year, while at Port Nolloth it is only two inches.

In Natal, on an average of years, rain falls at Pietermaritzburg on about 126 days in the year, of which 95 are in the six summer months, and 31 in the winter, while no rain falls on about 239 days. The amount of rainfall in the year varies considerably in different parts. The average rainfall in the winter months at Pietermaritzburg is 7 inches, and 30 inches in the summer. On the high lands above Ladysmith, the fall averages from 24 to 31 inches in the year. The rainfall at Durban, which is about 40 inches in the year on an average, and along the coast, is heavier than on the hill country in the interior. Although the total amount of rainfall in Natal is not very largely in excess of that in England, yet it appears far greater through being concentrated into a shorter season.

In Bechuanaland the winter—April to September—is dry; rain falls from November to April, which are the summer months. The water soon disappears, but it can be generally got by digging in or near the river beds, or by sinking wells. The annual rainfall is about 25 inches. "It has been remarked that the climate in this district is gradually becoming drier, but the lack of reliable meteorological observations renders it difficult to substantiate this statement. It appears, however, that in former years the hippopotamus was found in many pools of the Kuruman River, which are now nearly dry."

The climate of the Orange Free State is remarkable for its dryness, the rain, which averages about 22 inches yearly, falling principally during the violent thunderstorms which occur at frequent intervals during the summer. But the torrents of rain that then descend are soon absorbed by the dry ground, and are quickly carried off by the gaping sluits and deep river-beds that seam the grass-covered surface of the veldt.

The Transvaal is, on the whole, much better watered than the Free State. The winter season, as in all Eastern South Africa, is dry, the rains commencing in September, setting in heavily about November—frequently rendering the roads almost impassable—and ending in March or April. Thunderstorms and hailstorms are frequent during the summer months, while in winter the high veldt and the mountains are covered with snow. Mr. Miles, C.E., gives the total rainfall at Johannesburg, in 1890, at 25·94 inches, falling on six days; in 1891 it amounted to 40·85 inches, falling on 99 days; in 1892 it was 27·54 inches, falling on 94 days. In some years not a drop of rain has fallen for five months at a time. At Pretoria the rainfall seems to vary from 20 to 30 inches.

In the higher mountain districts all over South Africa snow falls during the winter; but there is not a single summit that reaches the limit of perpetual snow. In Basutoland, the winter months—June, July, August—are dry, rain seldom falling from May to October. Frequent showers fall in October, November, and December, but January and February are the rainy months. The rainfall is also abundant in the Chartered Company's territories—both south and north of the Zambesi. The low-lying coast-lands of Portuguese East Africa have a very heavy rainfall; but the German territory on the opposite side of the continent is almost rainless along the coast and in the interior towards the Kalahari, although the uplands are, on the whole, fairly well supplied with water.

In the eastern portion of South Africa, *hot winds* are occasionally experienced during the summer; they come from the north-west, carrying with them waves of heated air from the central plains, and blowing as if from a furnace. Fortunately, they are not of long duration anywhere. In some parts of Cape Colony, and also in the Transvaal, violent hailstorms occur, and do much damage to vegeta-

tion and stock. Thunderstorms are very rare in the neighbourhood of Cape Town, but in almost every other part of South Africa they are frequent in summer, and often very grand. Fleecy clouds rise on the horizon, swelling and darkening until the lightning flashes along them, while the thunder peals out with long and increasing reverberations. It is then a sight to watch the brilliant colours and forms of the electric discharges, and their varied track against the inky black sky—now forked, now straight, now zigzagged, now in quivering rays and horizontal flashes, appearing and disappearing rapidly in the twinkling of an eye. Such striking exhibitions of the forces of nature, however, do not last long ; after them the rain ceases, clouds roll up and disperse, and a delicious cool atmosphere follows.[*]

With the exception of the Portuguese coastlands, and the low-lying riverine districts along the Zambesi, South Africa may be said to be one of the healthiest and most salubrious countries in the world. This is amply proved by the fact that, in physique, the descendants of Europeans, both Dutch and English, have in no way deteriorated, but have in fact improved. The typical *Boer*—a tall, well-built, strong, healthy man—is a finer specimen of the genus *homo* than the present-day Hollander, while an Africander of English descent is a lithe, athletic, and sinewy fellow, much more active in his movements, and capable of undergoing greater hardships and severer privations, than an ordinary Englishman, even though he be a man of note on the football ground or on the cricket field. But besides the undoubtedly favourable effect of the climate on those born and bred in the country, it is "little less beneficial on those who, born in Europe, reside there. Children grow more rapidly and develop sooner, while a strong, sturdy manhood is followed by a happy old age, with fewer of the attendant senile diseases observed in the more vigorous climates of the north." As far as the native races are concerned, we

[*] *Official Handbook,* Colonial and Indian Exhibition, 1886.

may say that, while some of the lower types have decreased in number and will probably disappear altogether, the stronger Kaffir races show no signs of disappearing before the advance of the all-conquering European, but are increasing rapidly and undoubtedly improving, both physically and mentally, and seem destined in future generations to occupy a by no means ignoble position, industrially, intellectually, and politically, in the "United States of South Africa."

"*South Africa as a Health Resort*" is too wide a subject to be dealt with here, and mere generalisations are of but little value, especially when we consider the diversity of physical features and climatic conditions that are found in so extensive a country, and of which different parts are suitable for different ailments or for the different stages of disease, especially of the lungs, on which the dry invigorating air, particularly that of the interior plains and uplands, seems to exert the most beneficial influence. Dr. Lawrence Herman says that sufferers from bronchial and asthmatic affections derive great benefit from a stay in the country ; and that, where there is any marked hereditary phthisical tendency, a residence is most strongly to be recommended, particularly in the case of children. In other conditions of debilitating and wasting disease, much benefit will be derived from travelling in the equable climate of South Africa, and a convalescence from a serious illness can be most profitably spent by voyaging to the Cape, spending a short time there, and then returning.

The Voyage to the Cape has become justly famed as one of the most pleasant and enjoyable it is possible to make, and the fine ocean-steamers of the "Castle Line" have become celebrated for their comfort and punctuality. Each carries a surgeon, and is provided with everything necessary for the comfort and convenience of passengers during the voyage, which is accomplished in from 15 to 17 days, touching, outward and homeward, at Madeira or Grand Canary, and occasionally at St. Helena and Ascension.

IV.—THE SPORTSMAN'S PARADISE.

Nature will not bring back the mastodon—Nor will game laws renew the teeming animal life of the country- Lions round Cape Town—Plains literally darkened by all kinds of game—The interior still a "Sportsman's Paradise"—Hunting on the South African veldt—The matchless fauna of South Africa—The antelopes—The Cape buffalo—The zebra and quagga—The giraffe—The black and the white rhinoceros—The elephant—Beasts of prey—Hyænas, leopards and lions—Dangers of lion-hunting—Prolific bird life- Game birds of South Africa—Snakes and vipers—South African fishing—Lord Randolph Churchill's advice to young Englishmen—Indescribable charm of life on the veldt.

Nature will not bring back the mastodon, and neither will the game laws, passed within the last few years by the various governments in South Africa, renew the marvellous display of animal life that met the astonished gaze of the early settlers. When the Dutch landed in Table Bay, the country absolutely swarmed with animals of all kinds, packed, as Mr. Trimen says, with the *feræ naturæ* down to the very margin of the ocean. Lions then chased immense herds of "harts and elands" on the slopes of Table Mountain, and hippopotami crashed through reedy swamps on the very site of Cape Town.

Van Riebeck, in his diary, tells us that one night the lions—evidently desiring a change of diet—"appeared about to storm the fort for the sheep within it," and not long afterwards the king of beasts interviewed the Dutch Commander in his own garden. For fifty years lions prowled about the settlement, and as late as 1694 attacked a herd of cattle within sight of the castle. If the settlers ventured a few miles inland, they had sometimes to make a detour to avoid troops of elephants, while in the Paarl Valley were " many rhinoceroses and zebras, and great numbers of hippopotami in the river." In 1685, Com-

mander Van der Stel made his famous expedition to Namaqualand, and when somewhere in the vicinity of the Piquet Berg, we are told that a rhinoceros furiously charged his carriage and almost made an end of His Honour when he jumped out.*

Later explorers and naturalists also give us some idea of the matchless fauna of South Africa, but the innumerable hosts of animals with which the country teemed have been so ruthlessly hunted and so wantonly destroyed, that "never again will the traveller be able to stand upon his waggon box, and, like Burchell, Andrew Smith, Cornwallis Harris, and Gordon Cumming, scan plains literally darkened by thousands upon thousands of wildebeests, quaggas, Burchell's zebras, blesboks, hartebeests and springboks." Twenty-five years ago, the undulating plains of the Free State swarmed with game—over 25,000 head were driven before the guns of Prince Alfred's shooting party in 1860. Speaking of the Transvaal, Lord Randolph Churchill says that its wide and grassy plains formerly abounded with game of almost every description. Persons whose word can be implicitly relied on informed him that, within the last fifteen years, they remembered these plains being covered as far as the eye could reach with countless thousands of wildebeest, blesbok, springbok, and other varieties of the deer and antelope tribes. The Boers, however, slaughtered without discrimination every wild four-footed animal. "Forming themselves into large shooting parties, they shot the beasts down everywhere by scores, and by hundreds, and by thousands, leaving the carcases to be devoured by the vultures, and going a few days afterwards to gather up the skins which the vultures had neglected, and which the sun had dried and tanned."† Little wonder, then, that the traveller can now compass mile after mile of plain without seeing so much as a solitary buck.

But although all over the settled parts of South Africa the slaughter has been reckless and ruthless in the extreme, there are yet large areas where good shooting may be had, while many a tract in the more inaccessible regions is still a "sportman's paradise." Excellent sport is to be obtained in plenty by the sportsman who does not fear hard work, and has made up his mind to journey far into the interior, where many a gallant head of game and many a rare species will fall to a well-aimed rifle. In the Pungwe River hunting-grounds a prodigious quantity of game—big and little—swarms. Buffalo, hippopotamus, rhinoceros, buck of all kinds — neither wild nor wary—teem in the swamps and thickets on either side of the railway. In Mashonaland, game abounds. "Antelopes of all kinds are numerous. Sable antelope, wildebeest, hartebeest, eland ostrich—all can be found and chased, though good galloping horses will be necessary for success, while the presence of many lions offers an exciting variation to the bold and steady shot." Further west and north, towards the Kalahari and the Zambesi, the hunter may, in a few weeks, come across giraffe, hippopotamus, ostrich, eland, sable antelope, roan antelope,

koodoo, wildebeest, hartebeest, waterbuck, zebra, many kinds of small buck, wart-hog, hyæna, and jackal, and probably leopard and lion. Almost every day he will find game of some kind, and, without much exertion, will be able to supply his camp with fresh meat. Even if he does not venture north of the Zambesi, he will find a wide range of animals, from the lordly lion and the vicious buffalo to the dainty steinbok and the stealthy duyker, that will test his powers of endurance as well as his skill as a marksman. Of the nobler game, Selous, the last of the great hunters in Southern Africa, shot lions, giraffes, buffaloes, hippopotami, and rhinoceros ; and the very perfect collection of South African fauna in the Natural History Museum at Cape Town has been mainly contributed by him.

With good horses, well-trained dogs, and proper guns, hunting in the South African veldt is as pleasurable as it is beneficial. "The morning rides through the bush have an indescribable charm. The scenery, the fresh air, the bright sunshine, and the knowledge that one may at any moment come upon anything in the shape of game— from a lion or a giraffe down to a pig or a baboon—lend to these excursions a most exhilarating interest."* The weather, in the winter, is perfect ; the days being bright and warm with refreshing breezes, while the nights are cool and even frosty. A few weeks of active, energising, and yet restful life on the grass-covered veldt is immeasurably more health-giving than months full of *ennui* in the soft air of the Riviera, or a winter in the dust-laden atmosphere of Egypt.

Among the matchless fauna of South Africa, the most conspicuous in number and variety are the *antelopes*, of which there are no fewer than thirty-one species, ranging from the massive eland and the princely koodoo to the little klipspringer and the diminutive blue-buck. The *steinbok* and the *duyker* (literally *diver*)—so-called because, when alarmed, it "dives" into the thickest bush— are among the smallest and most beautiful of South African antelopes, and are found abundantly on the open plains and in hilly grounds of a bushy character from the Cape to the Zambesi, and apparently thrive as happily on the driest pastures of the "Doornstland" as on the greenest hill-slopes. The *klipspringer* is another agile, handsome little antelope, but is only found among the most solitary and rugged mountains, where it remains even when snow falls. The *blesbok* formerly scoured the plains in myriads, but they are now rare. They are preserved on a few farms in the Cape, and may occasionally be met with in the Republics. Small herds of *springbok* still take their wonderful leaps over waggon tracks and paths all over South Africa, but are very difficult to shoot, much more so than the gallant little *boschbok*, which, as its name implies, is found in bushy country. Bush-bucks are common in the wooded kloofs of the Eastern Provinces and the bush veldt of the Limpopo, and the northern rivers—in the thick reed beds of which are hidden the few survivors of the beautiful red and white *rietbok*. The *waterbuck*—the *kringnat* of the Boers—is a handsome animal, the horns of the bull being very fine ;

---

* Noble's *Official Handbook*.

† *Men, Mines, and Animals in South Africa*. (London : Sampson Low & Co.)

* *Men, Mines, and Animals in South Africa*.

but it is not now abundant, even along the Botletle and other streams in the far north, where also is found the *lechwe*, or lesser water-buck. The shy and swift *vaal*, or *grey rhebok*, is fairly plentiful on the higher grounds, and the elegant and beautifully coloured *rooi rhebok* on the lower grounds, even in the Cape and Natal. The *grys steinbok* is now very scarce. The undaunted *gemsbok*, with its terrible horns, will attack and vanquish even the lion. Small troops of this buck (which figures on the Cape coat of arms, and is in all probability the original of the unicorn — the two horns, when seen in profile, appearing as one) still scour the country to the north of the Orange. The *bontebok* may be found in the Transvaal and Bechuanaland, but is extinct in Natal and the Cape, with the exception of a few preserved on one or two farms near Cape Agulhas. A few of the *zwaart wildebeest*, or white-tailed gnu, which also figures on the Cape coat of arms, are preserved in Victoria West. The *blaauw wildebeest*, or brindled gnu, is now extinct south of the Vaal and Orange, but is still plentiful in Khama's country and along the Okavango and in the Mababe country, and thence to the Zambesi.

The stately *koodoo* is a magnificent animal, and by far the most plentiful of the larger antelopes. It stands as high as a mule, and is of a soft grey colour ; its face is beautifully marked with white, and it carries fine twisting horns from two to three feet long. The sportsman will find large numbers of koodoo in North Bechuanaland, and in the bush veldt to the west of the Limpopo, and along the Zambesi. On both banks of these rivers, and round Lake Ngami, and along the Okavango and Chobe, will also be found small troops of *situtunga*, or Speke's antelope, the "water koodoo" of the trek Boers. The *pookoo* is another rare water antelope, found near the confluence of the Chobe and the Zambesi. The *hartebeest*, a fine antelope, but shy and swift, and the bloom-glossed *tsesseby*—the "Zulu hartebeest" of the Boers—are rare, except in the Kalahari, and along the Zambesi towards Mashonaland. Extremes meet in the massive *eland*, the largest of the antelopes, weighing from 900 to 1,000 lbs., standing six feet at the withers, and the *blaauwbok*, the smallest of all African antelopes, scarcely bigger than a rabbit. Immense troops of elands formerly pastured on the plains, but large herds are now only met with on the North Kalahari and towards the Zambesi. But the antelopes which, of all others, the South African hunter covets, are the *roan antelope*, which is about as big as a fine Scotch stag and quite as graceful, and the *sable antelope*, a magnificent creature with long horns arching over his back. These beautiful antelopes, both roan and sable, are still abundant in Mashonaland, and may be found in the northern portion of Khama's country and in the Mababe veldt.

The Cape *buffalo* is perhaps even more dangerous to hunt than the lion, its "muscular development and massive horny front, backed by a temper of sullen ferocity, render it an antagonist not to be trifled with, and hunting literature teems with the accidents and hairbreadth escapes" incident on encounters with this terrible and cunning beast. When wounded, it will lie *perdu*, waiting until the hunter gets close and then charging suddenly

and swiftly, often with irresistible force and fatal results. South of the Limpopo, the buffalo is only found in the Tzitzikama and Knysna forests, and in the Addo Bush, where it may not be shot without express license from the Government. In the marshy grounds of the Pungwe River hundreds have been shot, and vast herds still roam through the fever-stricken lowlands of the Barotse country.

Perhaps the most beautiful and characteristic of all equine animals is the *zebra*, which still survives in some of the wilder mountains of the eastern provinces. *Burchell's zebra*, which some travellers miscall quagga, is not now met with, except in localities where water is to be found, between Palapye and the Zambesi. The true *quagga*, formerly so numerous on the Cape karroos and the Free State plains, has, like so many other beautiful animals, become extinct. Happily the *giraffe*, the "kameel" or camel of the Boers, a unique and most peculiar species, still exists in the safe retreat of the great Thirstland of the North Kalahari. Very large troops of this magnificent animal roam over the desolate and waterless region to the south of the Botletle river, and they are also found in considerable numbers in parts of Ovampoland, and along the south bank of the Chobe. There are only a few now left in the North Transvaal and Mashonaland, where not so long since they abounded.

The *hippopotamus*, the "zee koe" (sea cow) of the Boers, once so abundant in every river throughout the Cape, is now confined to the lower waters of the Orange and the East Coast rivers to the north of Natal. Montsioa, the old chief of the Barolongs, told Mr. Bryden that he remembered sea cows in the Molopo, a "river" which is now a mere chain of pools, but they have long since disappeared from Southern Bechuanaland, and must be sought for in Lake Ngami, and in the Chobe and the Zambesi, where they are very numerous.

Next to the elephant, the *rhinoceros* is the most gigantic of all existing land animals. The *black rhinoceros*, says Mr. Selous, is still very plentiful throughout a large tract of country along the southern bank of the Central Zambesi, as it doubtless is also in many parts of the interior to the north of that great river ; and it will be many years, perhaps centuries, before it is altogether exterminated. But its congener, the great *white rhinoceros*, which may be regarded as specially South African, it being unknown north of the Zambesi, is on the verge of extinction, if not already extinct, although, twenty years ago, it was common over an enormous extent of country in Central South Africa. Two specimens have recently been shot by Mr. R. T. Coryndon, a well-known South African hunter. Both varieties have two horns, and in the huge white or square-nosed rhinoceros the front horn attained an enormous length of between 4 and 5 feet.

Like the buffalo, the *elephant*, with the exception of the small herds preserved in the Knysna forests and the Addo Bush, has been exterminated in the Cape, Natal, and the Transvaal, but large herds still frequent the well-watered bush country of Northern Mashonaland, while beyond the Zambesi there are plenty of them up the Loangwa Valley. In Ngamiland and the Kalahari country the wild elephant is scarce ; in fact, the amount of

ivory obtained from the whole country to the south of the Zambesi is now very small. According to Mr. Selous, elephant-hunting is even more exciting than lion-hunting, and never can this noble game be beheld by the South African hunter without pursuing it—a pursuit that entails great hardships, fatigue, thirst, and exposure to the intense heat of a tropical sun.

Man has not been the only destroyer of the antelopes and other herbivora of South Africa ; hyænas, cheetahs, caracals, servals, hunting dogs, leopards and lions have found ample scope for their powers of destruction. The noxious *hyæna*, both brown and striped, is happily fast disappearing from the settled districts ; the *cheetah*, or hunting leopard—the "luipaard" of the Boers—yet chases the antelopes on the upland plains ; while the *leopard*, the "tiger" of South Africa, is still common in all parts of the country, and, like the hyæna, often proves destructive to the flocks and herds of the farmers, by whom they are poisoned or trapped in large numbers. Although a leopard seldom ventures to attack man, it is, when brought to bay or wounded, a most dangerous animal to deal with, as savage as a lion, and as agile as a cat.

All these animals afford good sport, not wholly free from danger, but it is as a lion hunter that the South African sportsman reaches the pinnacle of fame. Unlike the leopard, the *lion* has disappeared from the country to the south of the Vaal and the Orange, but it still exists in large numbers in the bushier parts of Ngamiland and in the regions between the Limpopo and the Zambesi. When riding out from "Lion Camp," Lord Randolph Churchill saw a "yellow animal, about as big as a small bullock, lolloping along through and over the grass," and soon the glade appeared to be alive with lions. There they were, trooping and trotting ahead like a lot of enormous dogs, great yellow objects in the grassy veldt. Similar sights may still be enjoyed by the ardent sportsman who ventures beyond the Limpopo, but successful lion-shooting requires not only a cool nerve and excellent marksmanship, but the sportsman must be mounted on a perfectly-trained fast horse, for often enough the hunter becomes the hunted, and then it is a race for life.

The dangers of lion-hunting are vividly shown by an adventure that befell Selous when out riding one day looking for game. One of his Kaffirs, he tells us, jumped on an ant-hill and called out, "A lion ! a lion !" "Where ?" "There, there, close in front of you, lying flat on the ground." Selous instantly saw him—a male lion, crouched perfectly flat, with his head on his outstretched paws, and certainly not more than 20 yards from him. He was too close to feel inclined to dismount, and did not care to do so, especially as he was riding a good shooting horse lent him by Lobengula. His horse, however, would not keep perfectly still, and as he was trying to get the sight on to the lion's nose below the eyes, he saw him draw in his forelegs, which had been stretched out, under his chest ; then his whole body quivered. Selous knew well what these signs portended, and that the lion was on the point of charging. Just at that moment the intrepid hunter fired, and made a very lucky shot ; touching the trigger just as the sight crossed

the lion's face, the bullet struck him exactly between the eyes. Death was, of course, instantaneous. On returning to his camp, Selous had another and much more exciting adventure with a lion, which he also shot. Wounded lions are extremely dangerous to approach, and Selous says that anyone who has not seen at close quarters the fierce light that scintillates from the eyes of a wounded lion, can hardly imagine its wondrous brilliancy and furious concentration, gleaming with all the savage fury of unutterable though impotent rage.[*]

As the nobler game disappears, more attention will be given by the sportsman to the prolific bird life of South Africa. An English fowler can have no conception of the diversity of feathered game that lies everywhere at hand. *Guinea-fowls* swarm in vast numbers all over the country ; the *francolins*—the "pheasants" and "partridges" of South Africa—are plentiful ; many kinds of *bustards* abound, including the great *kori bustard*, or *gompaauw*—the king of South African game birds—and the *koorhaan*, one of the best of sporting birds ; the striking and graceful *Kaffir crane ;* the curious *secretary-bird ; eagles* of many kinds, *hawks* and *falcons* in great variety, and *vultures*—the "aasvogel" of the Boers—with the true wild *ostrich*, small flocks of which still scour the open plains of the Kalahari and Ngamiland. To shoot game birds in South Africa, a decent pointer is a *sine qua non*, and the sportsman must also acquaint himself with the game laws, which are now more or less strictly enforced in the country.

Snakes are, unfortunately, plentiful in South Africa. The green *tree-snake* is a most active reptile, but probably harmless. The *cobra*, or hooded snake, is tolerably common on the eastern coastlands ; its movements are swift and its temper fierce, and its bite is quickly fatal to man and beast. Here also is found the huge *python* or rock snake, sometimes twenty feet in length. The alert and pugnacious *ringhals* is also venomous ; but the most repulsive and dreaded are the *puff-adder* and other vipers, the flattened-backwards, broadened head of which, has "an expression that, for concentrated malignity and dull ferocity, has no equal throughout nature."

South African fishing certainly cannot compare with European or American sport ; but the enthusiastic disciple of Walton can find fair sport with rod and line in almost all the rivers and pools throughout the country. Naturally, as Mr. Bryden points out, siluroid fish occur most frequently, capable, as they are, of supporting life even when nothing but a mudhole remains to them.

Space will not permit us to enter into any details as to the outfit, the employment of Masarwas or Bushmen and Kaffirs as trackers, and Dutch hunters as assistants in the hunting veldt, etc. The sportsman will find ample directions and most valuable "wrinkles" in the works of Selous, Churchill, Bryden, Nicolls, Holub, Baines ; and, in addition to some of these, he should take with him "The Art of Travel," by Francis Galton, and "Shifts and Expedients of Camp Life," by Baines and Lord.

We cannot, perhaps, close this very brief outline of

---

[*] *Travels and Adventures in South-Eastern Africa.* By F. C. Selous. (London : Rowland Ward).

sport in South Africa better than by the following extract from Lord Randolph Churchill's well-known letters from South Africa, republished under the title of "Men, Mines, and Animals in South Africa," by Messrs. Sampson Low & Co., Ltd. :—

"To the young Englishman fond of study, of riding, of a wild hunter's life, active, vigorous, healthy, and endowed with adequate fortune, those regions of South Africa which extend from the Limpopo to the Hunyani River offer a field for sport not to be equalled in any other part of the world. During the winter time—from May to September—the climate of this region is almost perfect, the risk of fever slight. The air of the veldt is invigorating, the scenery and surroundings attractive and various, the life of the hunter temperate and wholesome. This man, coming to these parts of South Africa, eager for sport, will experience little, if any, disappointment. Accompanied and guided by some good Dutch hunter, he will see, pursue, probably kill, every African wild animal, with the exception of the elephant, buffalo, and rhinoceros. These also may be obtained without difficulty, if one is not daunted by the remoteness of the districts near the Zambesi, by the real rough life incident on the absence of waggons and of all beasts of burden, owing to the existence of the tsetse-fly, and by hard walking exercise under the heat of a tropical sun. But in the vast territory defined above, the hunter may, without difficulty, surround and cheer himself with every species of comfort. Waggons drawn by oxen or by mules—the former are preferable —can penetrate to any part of the bush veldt ; tents, bedsteads, provisions of all kinds, can be carried with ease ; and even a young Pall Mall sybarite would acknowledge that there can be provided out here an inconceivable combination of sport and luxury. The soundest sleep at night, the best of appetites for every meal, the clear head, the cool nerve, the muscle and wind as perfect as after an autumn in the Highlands, are pleasures and delights which can be here experienced, and to which many of our London jeunesse dorée are almost strangers. All kinds of strange forest sights, all the beauties and many quaint freaks of nature, will charm the eye and exercise the mind.

"Nor is the exciting element of danger by any means altogether absent. The lion and the leopard are beasts to encounter which, successfully, requires skill, experience, and courage. Snakes of great venom, and some of great size, may not infrequently be met with ; falls from the horse, when galloping wildly through the bush or over the plain—such as even Leicestershire cannot rival —may occur constantly ; and should anyone imagine that antelope-hunting in South Africa is a tame, safe kind of amusement, three or four weeks' experience of it will easily undeceive him. Then the game. Such numbers, such variety, such beauty ! Nothing more wildly lovely can be imagined than the sight of a herd of roan antelope, or hartebeest, or zebra, galloping through the forest ; nothing more wildly exciting than the pursuit of such a herd ; sighting the game through the trees, sometimes obtaining a fair standing shot at moderate range ; then mounting your horse, loading as you gallop along, leaving him to pick his way as best he can among the trees, branches, roots, stones, and holes ; coming again within one hundred and fifty yards, not dismounting, but almost flinging yourself off your horse, and firing both barrels as rapidly and as accurately as you may ; then on again, over hill, river, and dale, until you and your steed are alike exhausted. Then the accompaniments, the framework, as it were, of the chase : the early start, the break of day, the cool morning air ; the return to camp, wearied, but pleased and excited, the bath, the evening meal— eaten with an appetite and a zest such as only an African hunter knows ; the camp fire, the pipe, the discussion of the day's sport, the hunter's stories and experiences, the plans for the morrow—no thoughts of rain or bad weather oppressing the mind : all this makes a combination and concentration of human joy which Paradise might with difficulty rival. Nor is this hunting life, when pursued for a few months or from time to time, a useless, a frivolous, or a stupid existence, especially when it is compared with the sort of idle, unprofitable passing of the time experienced from year to year by numbers of young Englishmen of fortune. Nature and all her ways can be observed and studied with advantage ; much knowledge of wild animals and wild men can be acquired by the observant, the intelligent sportsman ; languages may be learnt, habitudes and customs noticed and written about ; interesting persons are met with, excellent friendships are formed ; the mind and the body are seasoned, hardened, developed, by travel in a wild country ; all its many incidents, its rough and its smooth, its surprises, its difficulties, its adversities, and its perils ; and I hold this for certain that, in nine cases out of ten, a young Englishman, who has had six months of South African hunting life, will be a better fellow all round than he was before he started."

V.—THE PEOPLE OF SOUTH AFRICA.

Advantages of South Africa—A thinly-peopled and imperfectly developed country—European inhabitants—Boers and Britons—English and Dutch—Native races—The Kaffirs—The Amakosa, Amazulu, and Bechuana Tribes—The Kaffir Language—Hottentots and Bushmen.

South Africa, says Mr. John Noble, "has a most healthful climate, where cloudless skies, continuous sunshine, and dry air can be enjoyed to perfection. Its lands give scope for every kind of pastoral and agricultural occupation. Flocks of sheep, herds of cattle, and troops of horses feed entirely on its natural plants and grasses. Its soils are fertile, offering the most ample choice to the cultivator, and producing almost anything and everything grown in tropical or in temperate latitudes. Its mineral deposits are varied and abundant, and some seem well-

nigh inexhaustible. Its flora is one of the richest on the earth's surface. Its fauna embraces the most interesting and conspicuous forms of the animal kingdom, and its inland regions are still ' the sportsman's paradise.' It has settled European communities, some of whom have for successive generations been engaged in the pioneering work of colonisation ; and within its borders are native populations, amenable to civilising influences and capable of becoming an increasingly-important and valuable industrial element."

But with all its advantages and conditions, so favourable to rapid progress, this vast region—a region equal in extent to all the countries of Western and Central Europe taken together, and not less than ten times the size of Great Britain and Ireland—is very thinly peopled, the entire population only amounting to about $4\frac{1}{2}$ millions ; and of those the Europeans, or persons of European descent, scarcely number three-quarters of a million. That is to say, a country more than twenty times the size of England has considerably fewer inhabitants than London, and not so many white people as Liverpool. In England, there are now nearly 600 people to the square mile ; in South Africa, there are four ! Of white people there are, on an average, only three to every five square miles.

Of South Africa, therefore, it may be said, as Professor Seeley said of the Empire as a whole, that it is as yet but very thinly peopled, and very imperfectly developed ; a young country with millions of acres of virgin soil and mineral wealth as yet not half explored, with abundant room for multitudes of Englishmen, and with homesteads for them all, for the most part in a congenial climate, and out of the reach of enemies. Here, if anywhere, is ample space for the English race to expand and renew, under the most propitious auspices, the mighty youth of the mother country ; and towards this favoured land should be directed the tide of British labour and capital that continues to flow over and fructify foreign and inferior countries ; for here our emigrants would themselves thrive under their own fig-tree, and rear children with stout limbs and colour in their cheeks and a chance before them of a human existence.*

The European inhabitants are mainly the descendants of the early Dutch settlers and later British immigrants. The South African-born Dutch are scattered all over the country as sheep and cattle farmers ; a large number of English people are also settled on the land, but most of them are to be found in the towns and mining centres. The genuine South African Boers lead a solitary and patriarchal existence on their farms, many of which extend for miles, and include large areas of the richest arable land, of which, however, the Boer cultivates but a very small part. He is quite content if he has sufficient pasturage for his cattle and a little seed-earth for his corn ; and the more rapid roads to wealth, especially if they necessitate residence in a town or a prolonged stay at a mining centre, have no attractions for him. Lord Randolph Churchill, during his tour through South Africa, formed a very poor opinion of the Boer population gener-

ally. The Boer farmer, he says, personifies useless idleness, passing his day doing absolutely nothing beyond smoking and drinking coffee, perfectly uneducated, and proud that his children should grow up as ignorant, as uncultivated, and as hopelessly unprogressive as himself. Other writers admire the Boer as a model pioneer, and regard the trek-Boers as marvels of hardihood and cool courage in the face of apparently insuperable and overwhelming difficulties. Selous dwells upon the simple kindness and great hospitality for which the Boers have always been noted, and says that no people in the world are more genuinely kind and hospitable to strangers than the South African Dutch. He is also convinced that, in South Africa, the Dutch element will never be swamped by the English, as it has been in America. The South African Dutch are one of the most prolific races in the world, and very large families of from twelve to sixteen children are not uncommon. They have good natural qualities, and only want education to enable them to hold their own against the Englishmen and Scotchmen, the Germans and Jews, who now fill the towns, exploit the mines, and carry on the trade of the country.

The Cape having been settled by the Dutch, and for a century and a half an exclusively Dutch possession, the Boers would naturally regard the forcible annexation of their country by the English with anger and distrust ; and subsequent events, particularly the emancipation of the slaves, deepened the hostility of the Boers, and caused them to regard Englishmen generally with intense dislike —their hatred of everything English culminating at last in the voluntary exile of hundreds of families, who left their homes and farms in the Cape and trekked away into the then unknown interior, across the Orange and over the Drakensberg, braving every difficulty and danger, and even death, rather than submit to what they considered an unjustifiable interference with their personal liberties.

For years, even after the founding of the republics, this feeling permeated the rural Dutch-speaking population of South Africa ; and, on the annexation of the Transvaal in 1877, the old hostility against the English flamed out anew in open rebellion in that country, and in a very bitter feeling among the Dutch of the Cape and the Free State. The magnanimous restoration of independence to the Transvaal, especially after the deplorable reverses at Bronkhorst Spruit and Majuba Hill, won the confidence of the great body of the Dutch people throughout South Africa, and led them to regard the British Government and their English fellow-colonists with feelings of respect, friendship, and trust. And as they got to know each other better, Boer and Briton became still friendlier and more disposed to work together amicably, helping, not hindering, the development which was seen to be good alike for both. The spirit of the Dutch, as a people, is too much like the English spirit to allow them to work with the English on anything but equal terms.* The work of Boer and Briton in South Africa is the same ; and no false assumption of superiority or useless regret for past blunders should be permitted to wreck "the chances of that peaceable expansion which is the complement of con-

---

* Froude.

* President Kruger.

ciliation." And to the continuance of this auspicious state of goodwill and mutual trust, the genius of the Cape Prime Minister, Mr. Cecil Rhodes, not less than the generosity of the surrender after Majuba, has contributed. He has known how to secure the confidence of the Dutch farmer, and, at the same time, to retain the trust of the English settler, and he has shown in the daily practice of his government that their interests are entirely and absolutely common.*

Although the Europeans or people of European descent in South Africa are chiefly of Dutch or English origin, there is a considerable foreign element—principally French and German. A large number of Huguenots, or French Protestants, driven from their own country by the revocation of the Edict of Nantes, in 1685, sought refuge in Holland, and about one hundred families were sent out to South Africa, where they settled and devoted "their best energies to the cultivation of the vine and the mulberry, the making of wine, the distillation of brandy, the production of silk, and the development of agriculture and horticulture generally." In a few generations, however, these exiles blended with their Dutch neighbours, and ceased to speak their own language. It was nearly a quarter of a century after the surrender of the Cape to the English before Great Britain took any effectual steps to plant her own people in the colony. In 1820, several thousands of British immigrants landed on the shores of Algoa Bay and founded Port Elizabeth, thence spreading inland over the lands between the Fish and the Sunday rivers. From time to time, numbers of people from other countries have settled in various parts of South Africa, the most numerous being the Germans, whose thrifty habits and steady industry have proved of inestimable value to the country.

English is the language in general use all over South Africa, especially in the towns and mining districts. *Dutch* is, of course, the official language in the Transvaal and the Free State, and may also be used in the Cape Parliament, but the rude *patois*, known as "Cape Dutch," is now only used by the Boer farmers in the country districts, and all attempts to make this dialect the national language of South Africa have failed. The relative importance of the two languages may be inferred from the number of newspapers published in each of them. No less than 64 English and only 7 Dutch journals are published in the Cape ; 7 English and none in Dutch in Natal ; 2 English and none in Dutch in Bechuanaland ; 2 English and 1 Dutch in the Free State ; 19 English and 3 Dutch in the Transvaal : in all, 94 English papers and 11 Dutch. There are no statistics as to the actual number of English and Dutch people in the various states and colonies, but in the Western Province of the Cape most of the country people are Dutch, while in the Eastern Province and Natal the English predominate. In the Free State the bulk of the people are Dutch ; in the Transvaal, since the great rush to the goldfields, the Dutch have been outnumbered by the British, German, and other immigrants. Nearly all the Dutch people in the Cape can speak English, and English is also taught

* Lord Randolph Churchill.

in almost all the schools in the Free State and the Transvaal.

The heterogeneous native races in South Africa greatly outnumber the white population ; but, if we consider the size of the country, their numbers are by no means excessive, being only about three to the square mile. They include the Mixed Races, the Bantu tribes, the Hottentots, and the Bushmen. The *Mixed Races* are descendants of immigrants from Java, Ceylon, Madagascar, &c., and various natives, and form "a motley population of every gradation of colour, feature, and physique." These "Cape Boys," as they are called, are of great service as day labourers and domestic servants.

The *Kaffirs* are the most numerous and widely spread of all the native tribes of South Africa. They belong to the great Bantu family—a race that includes all the African tribes from the Cape to the Congo, and from the Atlantic to the Indian Ocean, excepting only the Hottentots and the Bushmen, who seem to be the remnant of an aboriginal population that was driven to the south-western corner of the continent by the great Bantu wave from the north.

The typical Kaffir is, physically and mentally, superior not only to the Hottentot and the Bushmen, but also to the true Negro, and is a finer specimen of the genus *homo* than any other African race. Tall and well-proportioned, strong, muscular, erect, haughty, fearless, but cruel and callous, though sensible to kindness and consideration, with intellectual qualities of no mean order ; such is the Bantu, who has bravely fought both Boer and Briton, but whose assegai and shield proved of little avail against the rifle and the cannon.

The *Ama-Kosa*, or Kaffirs proper, occupy the beautiful and fertile coastlands between the Cape and Natal. Here are located the Gcalekas, the Gaikas, the Tembus, the Pondos, and other Bantu tribes, all of whom are now under the direct control of the Cape Government. The *Ama-Zulu* include not only the well-known Zulus of Zululand, but also the Swazis, the Tongas, the Manikoos, the Matabeles, the Mashonas, and other kindred tribes from Natal to the Zambesi. The *Bechuanas* are the most numerous and perhaps the finest-looking, though not the most warlike, of the South African Kaffirs. The various Bechuana tribes, of which the most notable are the Bamangwato and the Batlapin, are all found within the limits of the British Crown Colony and Protectorate. The Ova-Herero and Damara and Ovampo tribes, in the German Protectorate and the south of Angola, are also pure Bantus or Kaffirs.

The Kaffir language, says Theal, is rich in words, musical and euphonious, and there is no difficulty whatever in expressing any idea in it. The three "clicks" in Kaffir have been adopted from the Hottentot, and are somewhat difficult to sound, but European children very quickly master all the intricacies of Kaffir pronunciation and style, and are soon able to speak Kaffir as fluently as their own mother tongue.

Kaffirs are no longer mere savages. Contact with Europeans for two centuries has resulted in an appreciable advance towards a state of civilisation and progress. Thousands of them have "acquired handicrafts, engaged

in industrial trades, and accumulated fixed property ; many of them may proudly point to churches and chapels that have arisen chiefly from their own efforts, where large congregations, neatly dressed and well behaved, now regularly assemble at the sound of the Sabbath bell." The marvellous progress of the Basutos, during the last decade, and the prosperous condition of the Bamangwato under their enlightened chief, Khama, a Christian and a gentleman in the highest sense, show that, under firm and wise guidance, the Kaffir is capable of development and of occupying, in the near future, a much higher position among the peoples of South Africa than he does at present.

The Kaffir tribes show no signs of disappearing before the white man ; on the contrary, they are rapidly increasing in number. For instance, in 1839 the highest estimate of Bantu population, between the Umzimvubu and the Tugela, was under 10,000. Sixty years later, there were nearly a million Kaffirs on the same ground. And from the various Kaffir locations swarms periodically migrate and occupy vacant places, so that the great uninhabited wastes, that every traveller of a century ago describes, are now teeming with human life. That the Bantu population, from the Limpopo to the sea, has trebled itself by natural increase alone within fifty years, is asserting what must be far below the real rate of growth.[*] The sanguinary wars of merciless tyrants, such as Tshaka and Moselekatse, involved the destruction of hundreds of thousands of natives ; and Lo Bengula is said to have slaughtered over a million of the Mashona and other tribes. Inter-tribal conflicts, witchcraft, and " smelling out,' also widened the ghastly gaps in the Kaffir tribes ; but, with the conquest of the Matabele and the annexation of Pondoland, all this terrible waste of life may be said to have ceased ; and in a few generations the numbers of these prolific polygamists will be enormous, and will furnish an inexhaustible supply of labour for the development of the country. Given sufficient inducement and fair treatment, the Kaffir will work, and that steadily and well ; thousands of them are now working in the Diamond Fields and in the gold mines, the coal mines, and on the railways ; while, on the farms and in the households, the labour power is Kaffir. Europeans are only employed in skilled work, or to direct and superintend the labour of the natives.

Unlike the Kaffirs, the *Hottentots* and the *Bushmen*— the aboriginal, and once the dominant, races of Southern Africa—are steadily diminishing in numbers, and in a few generations will probably be extinct. The pure Hottentot race is already extinct in the Cape ; of the 50,000 classed as Hottentots in the census of 1891, but few, if indeed any, are pure-bred, and the Koranna and

Namaqua Hottentots found north of the Orange River are probably also a mixed race. The Bushmen are evidently the remains of a great primeval pigmy race that once overspread Southern Africa, but was broken up by successive immigrations of Hottentots and Kaffirs, and its fragments driven into almost uninhabitable deserts and inaccessible mountain fastnesses. These wretched nomads were ruthlessly hunted by the early Dutch settlers, who shot them down like vermin ; and disease, the assegai, and the rifle exterminated many a tribe of both Bushmen and Hottentots. The Kaffirs killed every Bushman they could find, and were frequently at war with the Hottentots, with whom, however, a process of amalgamation went on along the frontier, so that, in some cases, Hottentot tribes became Kaffir clans. Hottentot women also became the slaves of Dutch Boers, and from them sprang the bastard race now known as *Griquas*. " The Cape Hottentots and the Griquas have certainly emerged from barbarism. They have lost their indigenous manners and usages, but from their ancient pastoral habits retain traditionally their passionate love for the ' beesties,' and for this reason are universally employed in all occupations connected with horses and cattle." The *Korannas* in Bechuanaland, and the *Namaquas* and the *Hill Damaras* in the German Protectorate, are all classed as Hottentots, but the Koranna and Namaqua exhibit Bantu characteristics, while the miserable Hill Damaras differ only in colour from the abject Bushmen, who, however, have some good points. . Selous says that, as trackers and assistants in the hunting veldt, the Masarwas, or Bushmen, are unrivalled, especially if they are half starved, for as soon as they get fat they become lazy and careless like dogs. He says that there is one faculty which the Bushmen possess in an extraordinary degree, and that is the sense which enables them to find their way, by day or night, through level pathless forests, where there are no landmarks whatever, to any point which they wish to reach, where they have ever been before.

The Bushman's language is, like the race, primeval in character, and is apparently "a collection of clicks modified by grunts." The Hottentot tongue, on the contrary, which is now, however, only spoken by a few wandering tribes, is more highly developed than even the Bantu, though both have adopted the click, presumably from the Bushman. Both Bushmen and Hottentots are poetical in their ideas, and have an extensive traditionary literature full of wonderful myths and curious fables. In the Bushmen, the artistic faculty was well developed ; in the caves they inhabited are to be found coloured drawings in clay and ochre of animals and men, and representatives of a mythological character relating to their customs and superstitions.[*]

---

* Theal's *History of the Republics of South Africa*, 1889, p. 404.

* Noble.

VI.—THE MAKERS OF SOUTH AFRICA.

The real "Makers" of South Africa—Its farmers, miners, and traders—South African farms—Ostrich farming—Land under cultivation—Irrigation—The model farm of the Transvaal—Viticulture—Tobacco, sugar, and tea planting—Mining—Diamonds and gold—Copper and silver—The South African coalfield—Iron ore—Lead, cobalt, and other minerals—The traders of South Africa—South Africa, a single trade area—Means of communication—Railways and Roads—Commercial position of South Africa—A customer, and not a competitor, of England.

The real "Makers" of South Africa are its farmers, its miners, and its traders; the men who pasture their flocks and herds on the upland plains and mountain slopes, or cultivate the valleys and watered lands; the men who delve deep in the earth for the gold and silver, the diamonds and coal; and the men who collect and forward the produce of the farm and the mine, and distribute the necessaries and luxuries drawn from other lands.

The *farmers* occupy a foremost position among the "Makers" of South Africa; their industry is the first in order of time, and is first also in order of importance to the well-being of the country. The great bulk of the native population and the majority of the European inhabitants of South Africa are settled on the land, and are occupied in the rearing of stock or the cultivation of the soil. The farmers are chiefly occupied in the rearing of sheep and goats, the breeding of cattle and horses, and the country is, on the whole, better suited for pastoral pursuits than for agricultural operations. South African farms generally are very large, and the farmers are, for the most part, owners of the land which they occupy. *Wool* has always been and still is a most important and staple source of wealth, and the grasses of the South African veldt and the pasture plants of the karroos are admirably suited for the growing of the finest wool. Over seventeen million sheep are pastured in the Cape, one million in Natal, seven millions in the Free State, and several millions in the Transvaal, Bechuanaland, etc. Millions of the beautiful Angora goats, which yield the valuable *mohair*, are also reared, chiefly in the Cape; and in time, with careful and intelligent supervision, South Africa might surpass Turkey as a mohair-producing country. Cattle are extremely numerous in almost every part of the country, but the formerly enormous demand for oxen for transport-riding has been greatly diminished by the opening of railways, previous to which, practically, the whole of the goods traffic was carried on by means of ox-waggons.

In the neighbourhood of the larger towns, or near the lines of railway, *dairy-farming* is very profitable; but very little has been done towards the cultivation of food for milch cows or other stock, the animals being left to depend entirely on the natural veldt, with the result that in times of drought they get into a very poor condition, and numbers die. Attempts are now being made, and to all appearances will be as successful as similar attempts in Australia, to develop an export trade in butter and other dairy produce; but the present supply of milk and butter, especially in dry seasons, falls short of the local demand, and prices often run very high.

Curiously enough, although the natives of South Africa, at the time of the discovery of the Cape, had cattle, sheep, goats, dogs, and poultry, and the country was the native home of three species of zebras, *horses* were un-

known, and were first introduced by the Dutch East India Company from Java. Others came from South America, and the stock was from time to time improved by pure Arabs and English thoroughbreds. Cape horses have, however, deteriorated, and at present there is no extensive demand for them, although a few are exported to India for army purposes. "Bony, high-withered, and goose-backed," though the unbeautiful Cape horse is, it is hardy and enduring, of indomitable pluck, and capable of standing hot and cold weather in the open, and keeping in good condition on the natural veldt.

Another peculiar and most important industry of the Cape is *ostrich-farming*. Up to 1864, ostrich feathers were obtained only from the wild birds, and European and native hunters chased and killed them at all times of the year, until they were almost exterminated. Between 1857 and 1864, however, a few farmers had succeeded in rearing a number of wild ostrich chicks, and ostrich-farming soon became a recognised industry. No great advance, however, was made until Mr. Douglass perfected his incubator. Artificial hatching entailed artificial rearing, and thus the formerly wild and excessively shy bird has become perfectly domesticated. Ostriches are now bred and reared like poultry, but the price of the feathers fluctuates so much that only farmers with considerable capital and special knowledge and experience can succeed in an industry so dependent on the capricious whims of fashion.[*] There are now a quarter of a million ostriches in the Cape, and the industry is spreading to the north of the Orange and the Vaal. In order to preserve its monopoly, the Cape Government has imposed a tax of £100 on every ostrich, and £5 on every ostrich egg, exported. A graphic account of ostrich farming will be found in Mr. Douglass's paper in the *Official Handbook to the Cape*, and in Mrs. Annie Martin's brightly-written "*Home Life on an Ostrich Farm.*"

The land under actual cultivation, even in the long-settled districts of the Cape, is comparatively limited, although the agricultural capabilities of the country are practically inexhaustible. The rainfall in the south-western and eastern coast regions is ample; elsewhere, except in some favoured localities, it is impossible to rely upon a regular return from the soil without *irrigation*, which has many advantages, as "it enables more valuable crops to be grown than those which can be produced without irrigation, and in a climate like that of South Africa, which is warm enough for vegetation all the year round, it permits crops to be raised almost without

* For instance, though 259,933 lbs. of ostrich feathers were exported in 1893, as against 253,954 lbs. in 1892, yet their declared value was only £461,552, as against £1,093,939 in the former year.

interruption—one crop following another in close succession." *

The authorities in the Cape are doing their utmost to encourage irrigation—the largest irrigation work yet undertaken being *Van Wyk's Vlei* in the Carnarvon district. Where the rainfall is sufficient, splendid crops of the finest wheat, barley, and oats are grown; but maize, or "mealies," is a much more certain crop, and is more widely grown, forming, as it does, the staple food of the natives, from the Cape to the Zambesi. Kaffir corn, or millet, is largely grown for making native beer. Basutoland has perhaps the best wheat-growing land in all Africa, and the south-eastern districts of the Free State and the Transvaal are admirably adapted for the growth of the king of cereals.

Until recent years, however, the farmers north of the Orange did little in the way of cultivating the soil; and even now, with all the opportunities for profitable cultivation consequent on the settlement of a large mining population, the Boer farmer is very loth to exert himself. A well-to-do Boer was one day boasting that he had obtained exactly double the price which he had expected for his wheat. "I suppose," an English friend said, congratulating him, "that you will sow a double quantity this year." "A double quantity?" replied the Dutchman, "half the quantity you mean? Don't you see that, with a double price, half the quantity will give me the same return?" The description given by *The Times* Special Correspondent, in her "Letters from South Africa,"† of the Irene Estate, on the railway near Pretoria, shows what may be accomplished by intelligence and enterprise. At this place scientific farming has only been attempted for two years; and yet the writer says that if she were to endeavour to describe the full result, she should probably be accused of wishing to re-edit "Robinson Crusoe." "Everything that is written of the material resources of this astonishing country must read like exaggeration, and yet exaggeration is hardly possible. The fertility of the soil is no less amazing than the mineral wealth. Sowing and reaping go on all the year side by side, and there is no fallow time for the ground. Here were pea-nuts ready for reaping, and green oats, barley in the ear and barley in the shoot, Swedish turnips fit for storing and Swedish turnips just shooting, mangold-wurzel, also in both stages, rye in the ear, carrots quite young and carrots ready for storing, potatoes in both stages; and in one immense field the sowers and the reapers had literally met. At the far end maize was standing, reapers were busy cutting and carrying the sheaves of corn; upon their heels sowers followed putting the wheat into the ground; and at the near end, where maize had been standing ten days before, thin green blades of wheat were already shooting. So vigorous is the growth of everything, that forest trees planted only two years ago were already high enough to give shade; apples grown from seed of March, and grafted in October, will bear fruit this year. With the exception of cherries, gooseberries, and currants, all European fruits

flourish well. Throughout the estate, the watercourses which divided the fields were bordered by hedges of quince, pear, apple, plum, and peach. The gardens contained a profusion of European vegetables and fruit-trees. Acres of roses, violets, and ornamental plants surrounded the houses: but nothing seemed to impress upon me more vividly the rapidity with which the place had sprung into being than the simple fact that, after hours of driving through vineyards, woods, and cornfields, we were met at the door of the house by a baby child of two and a half, who was older than everything we had seen. The estate had been named after her. When she was born, the spot on which it stands was nothing but bare veldt." And in almost every other part of South Africa, where the rains are copious, or irrigation possible, the soil and climate are equally favourable, and only require capital and energy to be brought to bear upon them to yield magnificent results.

But the extraordinary fertility of the soil in South Africa is shown, perhaps, most strikingly by the Cape vineyards, the produce of which in quantity and quality surpass those of any other vine-producing country in the world. In the coast districts of the Cape, the average yield amounts to 86½ hectolitres per 10,000 vines; in the inland districts it is, on an average, 173 hectolitres; but many farmers in the Worcester, Montagu, and Ladismith districts, obtain, year after year, as much as 5 leaguers from 1,000 vines, which amounts to what a European wine-farmer would consider the incredible quantity of 267 hectolitres per 10,000 vines! Now, according to Baron von Babo, the greatest living authority on viticulture, the average production of wine per hectare of 10,000 vines in Italy, in the United States, and in Australia, is 14½ hectolitres; in Spain 17, in Greece 17⅔, in France 18¼, Austria 18½, Hungary and Germany 24, Algeria 25½, and Switzerland 42 hectolitres; that is, at the Cape, ten to twenty times the amount of wine can be raised as from the same area in Australia or the States, and twice to six times the amount obtained in Switzerland. Unfortunately, the Phylloxera made its appearance in the Cape in 1886, and, in spite of all attempts to eradicate this dreaded pest, it has spread rapidly; but nurseries have been established by the Government for the purpose of raising such varieties of American vines as have been proved to be phylloxera-resisting. The varieties of the Cape grapes, grafted on these American vines, are much larger than from vines which were planted directly from the cuttings, and this augurs a bright future for the Cape as a wine-producing country. The Constantia and a few other wines have a deservedly high reputation, but Cape wine and brandy generally are of inferior quality, and are consumed principally by the natives; and it is "a sign of the robustness and vitality of the indigenous races in South Africa that they have not yet been exterminated by 'Cape Smoke.'"

Tobacco is also cultivated in the Cape and Natal and in the Transvaal, but the industry is still in its embryo state, although capable of development. *Sugar* and *tea* planting in Natal have passed beyond the experimental stage; a considerable quantity of sugar of excellent quality is exported, and the tea produced is of good

---

* Chisholm.

† Republished in book-form by Messrs. Macmillan & Co., London and New York.

flavour. *Coffee* and *arrowroot* also thrive on the moist coastlands. No greater stimulus could be given to the development of the sugar, tea, tobacco, and other industries in South Africa, than the free exchange of the indigenous products of the various colonies and states, and the formation of a Customs Union for the whole of South Africa, a region which, from a practical point of view, forms but a single trade area.

Mining has become so prominent an industry in South Africa that, judged simply by the value of their products, the miners already rival, and may before long surpass, the farmers in importance as "Makers" of the country. The mineral wealth of South Africa is amazing ; its stores of diamonds and gold are practically inexhaustible ; while abundant supplies of coal and iron will quicken its industrial development, and put its progress upon a permanent and stable basis. The wealth of the country in diamonds and gold has been already referred to ; those who desire further information should read Mr. Theodore Reunert's exhaustive and graphic work—*Diamonds and Gold in South Africa.* The shutting down of important diamond mines, and the restriction of the output in order to keep up the price,* has made gold-mining the most important of all South African industries, but the country contains rich deposits of other metals and minerals, but few of which have as yet been worked.

The *copper* mines of Namaqualand are not surpassed in richness of yield by those of any other country ; and the new *silver* mines in the Transvaal may, in the near future, prove no mean rival to the gold-fields. The principal silver mine now worked is in the district of Pretoria, about 50 miles east of Johannesburg, and six miles from the coal-fields, upon which the silver industry may be said to be entirely dependent for its existence. Some samples of ore from this mine run over a thousand ounces to the ton, and the reputedly argentiferous country is some thousands of square miles in extent.

Next to the precious metals and diamonds, the future of South Africa rests upon its *coal* and *iron.* The great coalfield of South Africa embraces an area of some 56,000 square miles, and extends from Burghersdorp to Aliwal North in the Cape Colony, and thence to the vicinity of Bloemfontein in the Free State, Heidelberg and Lake Chrissie in the Transvaal, there bending south-eastward to Newcastle and Ladysmith in Natal, and along the eastern foot of the Drakensberg to the Stormberg, above

Queenstown. The principal coal-mines in the Cape are at Cyphergat, Molteno, Fair View, and Indwe. The Indwe Mine, about 60 miles north-east of Queenstown, is considered the centre of the Cape coal area. The Dundee coalfield, in Natal, now has an annual output of over 100,000 tons of coal, adapted for general steam purposes as well as for domestic use, while, in some districts in the Transvaal, the deposits are so numerous as practically to form a continuous coal-bed over a large area of country ; in most cases the main seam is of considerable thickness, in many places being over 20 feet thick, ten feet thickness of clean coal being very common.* A very important colliery is now in full work at Vereeniging, on the Transvaal-Free State Railway, near the junction of the Vaal with the Klip river, and about 30 miles south of Johannesburg. An outlying deposit is also being worked at Boksburg, about twelve miles east of Johannesburg, and at Brakspan and the Springs, to the east of Boksburg ; but the most extensive deposits are those of the Oliphants and Wilge Rivers district, through which the Delagoa Bay-Pretoria Railway will pass. "The proximity of large beds of coal to the goldfields on the Rand has b'en of immense value in their development, and but for this singular and most happy juxtaposition of the coal and the gold, many of the mines would not be worth working at all. And as railway communication is opened up between the De Kaap and other goldfields and the collieries, we may expect an enormously greater production of the precious metal."

Another fact, pregnant with most important consequences to the future of South Africa, is that "in close proximity with the coal there are enormous deposits of the finest *iron* ore, especially in Natal and the Transvaal. In both countries, the natives have for years extracted and used the metal for the manufacture of battle-axes, assegais, and other weapons. *Lead* ore, with an unusual proportion of silver, abounds in the Transvaal, especially in the Marico district, and cobalt is found in the Middelburg district, while rich deposits of tin have been discovered in Swaziland. *Crocidolite* occurs in Griqualand West, and as for the *diamond,* the Kimberley mines may find formidable rivals in the Free State and the Transvaal. A new *diamond* mine is being opened near Kroonstadt in the Free State, and the gem has been discovered on the banks of the Crocodile River and elsewhere in the sister republic. Besides these, *platinum* and *plumbago, manganese* and the *garnet, agate, amethyst, jasper, chalcedony* and other precious stones, *marble* equal to the best Carrara, *building-stone* and *lime,* occur in various parts of the country. As a mining country, then, Africa south of the Zambesi has certainly a brilliant future, and where the miners go, the farmers follow, and thus districts, that might otherwise remain unoccupied for generations, save perhaps by a native clan or two or a few half nomad Boers, become peopled and endowed with all the comforts and conveniences of an advanced civilisation in a few years.

The *traders* of South Africa, although they are in no sense producers, like the farmers and the miners, but

---

* "The world's stock of diamonds has increased enormously during the past fifteen years. In 1876, the output of the South African mines was about 1,500,000 carats ; in 1893, it was over 4,000,000 carats, and the great trust, which controls all the principal mines, asserts that it has 16,000,000 carats in hand at the present time. Meantime, the demand for diamonds has greatly increased, and they are more expensive to-day—partly because of the trust, and partly because of the increased demand—than they were a short time ago.

In one respect the diamond industry is different from almost all others. Its product, that is of gems, is never consumed. Of gold and silver, a much larger amount than most people would believe is literally consumed in the arts past recovery, but a diamond once cut goes into the world's great stock, and is liable to come upon the market at any time."

* Ernest Williams, Esq., M.I.C.E.

simply intermediaries between producer and consumer, yet rank high among the "makers" of the country, and especially of such a country as South Africa. The great merchants of the ports and inland towns, and the village or wayside storekeepers, have been very actively engaged in the "making" of the country, and a large amount of business has been and is being done by the travelling trader—the "trader" *par excellence*—the man who loads up his waggon with goods likely to tempt the natives, and fearlessly treks from tribe to tribe, returning to town or port, after an absence of many months, to dispose of the ivory, horns, skins, or feathers that he has received in exchange for his wares, and to renew his supplies for another trip. The hardy and resolute pioneer traders of South Africa have been the real discoverers of the country. The trader has always preceded the settler, and his depôt has been, as it were, an outpost of civilisation, which sooner or later became a centre of settlement. As in the past, so now, the track of a solitary trader's waggon across the pathless veldt, often becomes a highway for miner and farmer to advance still further into the heart of Inner South Africa.

Commercially, South Africa is but a *single trade area*, with a gathering ground of over a million and a quarter square miles, and a seaboard of over three thousand miles in length, in which there are numerous outlets and inlets for the external trade of the country. The collective commerce of this vast and homogeneous region is technically termed "the Cape Trade," and is carried on principally by the Castle and the Union lines of steamers—their fast and powerful mail steamers and intermediate boats giving practically a semi-weekly service between England and South and South-East Africa, and fortnightly sailings from the Continent.

The *means of communication*, external and internal, are excellent. Besides the mail steamers, which maintain regular communication with Europe and Australia, numbers of sailing and steam vessels, from various parts of the world, are found in the ports, from the more important of which railways penetrate the country, and are being extended and inter-connected, so that in a few years every important centre of population and settlement between the Cape and the Zambesi will be easily accessible. There is now through communication from Cape Town, Port Elizabeth, and East London, *via* Kimberley, to Mafeking in Bechuanaland, and through Bloemfontein in the Free State to Johannesburg and Pretoria in the Transvaal. The Cape-Pretoria trunk line is over a thousand miles in length, and the journey occupies about 60 hours. The Natal main line has been completed to Charlestown on the Transvaal frontier, within 120 miles of Johannesburg, and a survey is now being made for extending the line to that busy centre of goldmining. One branch of the Natal railway runs from Biggarsberg to the Dundee coalfield; another branch winds through Van Reenen's Pass, in the Drakensberg Mountains, to Harrismith, whence it will be extended to meet the main line from the Cape at Kroonstad. The Delagoa Bay-Pretoria line is rapidly approaching completion; a section of it is already open, and branch lines are being made to connect with the De Kaap Goldfields

to the south and the Murchison Goldfields to the north of the main line.. The Beira Railway is open from Fontesvilla, at the head of navigation on the Pungwe, as far as Chimoio, at the foot of the Manica plateau, and will soon emerge on the Mashona uplands—the objective being Salisbury, the capital of Mashonaland.

Besides the railways, there are, in the more settled districts of South Africa, fairly good *roads* on all the main lines of traffic, with substantial bridges across nearly all the larger rivers. From the various stations on and termini of the railways, *coaches* and *mail carts* convey passengers, parcels, and mails; heavy goods being forwarded to their destination chiefly by ox-waggons. Formerly, the lumbering ox-waggon, with its white canvas tent and long team of oxen, was the peculiar "institution" of South Africa; it was the "ship" of the veldt and the karroo. With their ox-waggons, the disaffected Boers of the Cape "trekked" north to escape from British control. By drawing them together in a circle or hollow square, and filling up the openings with thorny bushes, they formed a "waggon-laager" or entrenched camp, which often enabled them to check the onslaught of the savage hordes that assailed them. The ox-waggon has, in fact, been the means by which nearly all the pioneer work of colonisation has been done in South Africa; and in the development of all pastoral and agricultural pursuits, the ox and the waggon are still essential elements.

The *commercial position* of South Africa is also excellent in every respect. Before the opening of the Suez Canal, the Cape was the "Halfway House" between Europe and the East; and were the so-called "Overland Route" to India to be closed or even imperilled, Table Bay and Simon's Bay would immediately regain far more than their former importance as the chief commercial and strategical points on the only alternative ocean-route. Dr. Yeats, in his *Map Studies of the Mercantile World*,[*] shows very clearly why the commercial position of the Cape, and South Africa generally, is so favourable, and how it is that the country is a customer, and not a competitor, of England.

Regular sea communication with Europe; direct routes to Australasia and the East on the one hand, and America on the other; a lengthened sea-board, including several safe and commodious ports; a pastoral, non-manufacturing Boer population, with an ever-increasing influx of energetic and intelligent British colonists, who bring their advanced home-country knowledge to bear on their pursuits, and who are extending communication by railways and improved roads, so that produce of all kinds can come forward with more ease; all these combine to create and increase very considerable commercial movements. "The gravitation and circulation of goods, as throughout Africa, is to and from the sea-coast. There is but little trade between town and town, all being supplied from the great seaport centres. The duty of our merchants and traders is to watch the advance and extension of the railways and the increase of the towns, and be ready to

* Published by Messrs. George Philip & Son, London and Liverpool.

supply the well-known wants of the people." The "Cape Trade" is an increasing one, the circulation of goods yearly becoming of greater value—the annual import and export trade of the region south of the Zambesi now being not far short of thirty millions sterling, and as the greater part of this trade is with the mother-country, the commerce of South Africa is a very important item in British trade returns.

The position of South Africa is also, as Dr. Yeats points out, unique, in that it is a *supplier* and a *customer*, and not a *competitor*, of England. Generally speaking, the people are thinly scattered over extensive territories, and, "turning their attention to the land, to the improvement of sheep and cattle runs, ostrich farms, etc., do not attempt *industrial* life in many forms; and even where a large industrial population is concentrated, as at Kimberley and Johannesburg, their labour and capital are devoted to the natural 'earth-gifts,' and not to the production of commodities which would displace or render unnecessary the import of products 'made in England.' The people of the Diamond Fields and the Gold Fields, and the scattered Dutch as well as the English farmers and traders, *look to England to supply their requirements.*"

By the splendid vessels of the Castle Line, and by other vessels, South Africa sends to England her multifarious products, receiving in exchange every manufactured article required in the "opening out" and development of the country. We reiterate the fact that the people of South Africa are *customers*, and not *competitors*, in the hope that greater attention will be paid by our merchants and manufacturers to their special requirements. Other markets are becoming closed to us by the extension of native industries, or are filched from us by the ubiquitous Teuton or the smart American, who show a far greater readiness than we do to adapt their products to the special wants, or it may be the whims, of the various markets. "Trade follows the flag," and our merchants and manufacturers have an advantage over their foreign rivals in the common sentiments and mutual interests, which, in spite of all misunderstandings and mistakes, yet bind Britons abroad and Britons at home. But this natural preference is, as Dr. Yeats justly remarks, of no avail, if British goods are dearer or of poorer quality than foreign goods. The increasingly aggressive competition of Germany and the United States is, however, making itself felt even in South Africa, but the prospects of British trade in this rising country are bright, and indeed brilliant. The new life and energy which the discovery and development of the Diamond Fields gave to the entire country has been intensified by the discovery of the richest goldfields in the world. "Railways are being pushed forward, the population is increasing, the tide of emigration is setting steadily in this direction, and the prospect of enlarged trade with the Cape and Natal is unsurpassed, *because there is practically the whole of the African continent before them.*" The riches of the interior will be tapped from the south and south-east, and trade and civilisation will steadily advance north. The gate to the heart of Africa is not through Egypt, but through South Africa, and it is within the bounds of possibility that, even in our time, the valleys and plains of Inner Africa will vibrate with the tread of the iron horse, and that the trans-continental railway from the Cape to Cairo will be an accomplished fact.

VII.—THE STORY OF SOUTH AFRICA.

The Portuguese navigators of the fifteenth century—Prince Henry, the navigator—Phœnician circumnavigators of Africa—Diego Cam—Bartholomew Diaz—El Cabo de todos tormentos—El Cabo de Bea Esperanza—Pedrão Corvilhão—Vasco da Gama and the seaway to India—Death of the Viceroy d'Almeida—Sir Francis Drake—Dutch and English ships at the Cape—The Netherlands East India Company—Arrival of Van Riebeck at Table Bay—The first true colonists of South Africa—War with the Hottentots—Purchase of territory—Governor van der Stell's "haughty and unrighteous tyranny"—Origin of trekking—Revocation of the Edict of Nantes, and arrival of the Huguenots—Hottentots and Bushmen—The Kaffirs—The first Kaffir War—End of the Dutch East India Company's rule—English and French fleets at the Cape—The second Kaffir War—Revolt of burghers—Tiny republics—Surrender of the Colony to the English—Temporary British occupation—Third Kaffir War—Re-occupation of the Cape by the Dutch—General Janssens capitulates—The Cape again a British colony—Slaves—Boers and natives—The fourth Kaffir War—Important concessions—Slagter's Nek—The fifth Kaffir War—The British settlers of 1820—Magna Charta of the natives—Emancipation of the slaves—The sixth Kaffir War—The Earl of Glenelg's policy—Wars and devastations of Tshaka—Natal, a black Arcadia—Moselekatse—The Bechuanas—Moshesh—The Great Trek—The Matabeles—Gazaland—Relief and Dingaan—Massacre of the emigrants—"Dingaan's Day"—Panda—Republic of Natalia"—Orange River sovereignty—The Cape girdled by native treaty states—The seventh Kaffir War—Sir Harry Smith—The Boers defeated at Boomplaats—Sand River Convention—Establishment of the Orange Free State—The Cape Constitution—Anti-convict agitation—Sir George Grey's policy—The eighth Kaffir War—The wreck of the *Birkenhead*—The cattle-killing mania—British Kaffraria—The Transkeian territories—The ninth Kaffir War—Walfish Bay—The German Protectorate—Natal and Zululand—Rebellion of Langalibalele—Cetewayo—Isandlwana and Ulundi—The Orange Free State and Basutoland—Discovery of diamonds—The South African Republic—Annexation of the Transvaal—War of Independence—Majuba and Lang's Nek—Bechuanaland—Zambesia and the Chartered Company—The Pioneer Expedition into Mashonaland—The Matabele War—The Nyasaland Protectorate—Portuguese East Africa.

The Portuguese navigators of the fifteenth century were the hardiest and most daring of all seafaring peoples of Europe; and Prince Henry, the son of King John II. of Portugal and Philippa of Lancaster, sister of Henry IV. of England, was possessed with an insatiable passion for unravelling the mysteries of the unknown seas. Having accompanied his father on an expedition against the Moors in North Africa, his interest was centred on the Dark Continent, the southern limits of which were then unknown, but which, according to Herodotus, had been circumnavigated by an Egyptian fleet manned by Phœnicians, about six hundred years before the birth of Christ. These

C

adventurous mariners had sailed south from the Red Sea, and had returned to Egypt through the "Pillars of Hercules," reporting that "in sailing round Africa, they had the sun on their right hand." But this and other traditional voyages, if ever accomplished, had left nothing but vague legends, and although it is known that, in the first century of the Christian era, Arab sailors had rounded the "Eastern Horn" of Africa and had crept down the coast as far south as Quilon, if not Sofala, in their quest for gold, the southern extremity of the continent remained unknown to the civilised world until an intrepid band of navigators from Portugal, in their eager and persistent search for an open ocean-route to India and the East, sailed from point to point along the low and deadly West Coast, and finally doubled the long-sought headland and entered the Eastern Seas by way of the South Atlantic.

It was in 1484, eight years before Columbus set out to discover a western route to the Indies, that Diego Cam reached the mouth of the Congo, and, penetrating south, landed near Walfish Bay, and erected a cross on the headland now known as Cape Cross. Two years later, two little caravels and a small store-ship left Lisbon under the command of Bartholomew Diaz, and sailed still further south, anchoring in the 'little bay' of Angra Pequeña. Having set up a cross as a mark of possession, the little fleet proceeded on its voyage; but when off the mouth of the Orange, the vessels were caught in a furious gale, and for thirteen days were driven helplessly before it far to the southward past the Cape. When the storm abated, Diaz sailed to the east; but finding no land, as he expected, he took a northerly course, and made the land somewhere between Cape Agulhas and the Knysna. Sailing eastwards along the coast, the vessel entered Algoa Bay. Diaz landed on a surf-beaten islet in the bay, and erected thereon a pillar and a cross—those twin emblems of civilisation and Christianity. He greatly desired to proceed, but the crews complained that they were worn out with fatigue, and that behind them was some great cape, and that they had better turn back to seek it. Diaz persuaded them to press on a couple of days longer, promising to return if they did not by that time make some discovery that would induce them to continue. No such discovery was made, and when off the mouth of the Great Fish River the vessels were put about, Diaz sighting on the way home a bold headland, which he named "El Cabo de todos tormentos"—"The Cape of all the Storms," a name of ill-omen that King John, believing that its discovery gave "good hope" of an open ocean route to India, changed into the more auspicious one of "El Cabo de Boa Esperanza"—"The Cape of Good Hope."

While Diaz was thus engaged, King John had sent Pedrão Corvilhão to gather all the information he could about the East. Corvilhão embarked at Aden in an Arab vessel bound for Calcutta and Goa, and thence he crossed to the African coast and managed to reach Sofala. His messengers reported that vessels, sailing south from Portugal, would certainly reach the extremity of the African continent, and would thus arrive in the Eastern Seas. These encouraging discoveries determined Emmanuel, the successor of King John, to make another bold attempt to reach India by sea ; and in 1497, the famous Vasco da

Gama started on that adventurous voyage which forms the subject of Camoens' great national epic, "The Lusiad." From the Tagus, Da Gama put to sea in command of four small vessels, which, five and a half months later, anchored in St. Helena Bay, about 120 miles to the north of the Cape of Good Hope, which was shortly after doubled. After touching at what is now called Mossel Bay, Da Gama sailed along the coast, and on Christmas Day, 1497, sighted the bold headland and wooded coastlands of Natal. Thence the expedition went north, touching, en route, at Delagoa Bay, Sofala, Mozambique, and Melinde, where Da Gama engaged Arab pilots to take the vessels to Calicut, on the Malabar coast of India. This famous expedition returned home by way of the Cape, and arrived at Lisbon in September, 1499, after an absence of little more than two years.

The seaway to India and the East was now open, and an immense trade was carried on by means of the fleets which every year sailed to and from Portugal. In 1503, Antonio da Saldanha entered a bay which had never been entered before, and climbed to the summit of a great flat-topped mass of rock, to which he gave the name of Table Mountain. The bay, in which he anchored, was thereafter called after him the watering-place of Saldanha, until nearly a century later it received from the Dutch sea-captain, Joris Van Spilbergen, its present name of Table Bay.* Seven years after, Francisco d'Almeida, the first Viceroy of the Portuguese possessions in the East, on his return from India, put into Table Bay for water, and, having seized some cattle, was attacked by Hottentots and killed, together with sixty-five of his men. After this, the Portuguese avoided the Cape as much as possible, and although for more than a century their fleets passed and repassed the Cape, year after year, they seldom touched at any port south of Sofala ; in fact, the Portuguese did little more than discover South Africa.

In 1580, the gallant Sir Francis Drake, on his famous voyage round the world, sighted the Cape, which he describes as "a most stately thing, and the fairest cape we saw in the whole circumference of the world." The English flag was first seen in Table Bay at the end of July, 1591, when three ships touched at the port on their way to India. The Dutch made their first appearance at the Cape in 1595, and in 1598 the Dutch ship Lion called at Table Bay with John Davis, the famous Arctic and East India Navigator, on board.† Three years later, the first fleet of the London East India Company put into the bay, and successive fleets of the same company also made Table Bay a port of call for water and refreshment.

The over-sea trade with the East soon proved so lucrative that several companies were formed in the Netherlands to profit by it ; but, to avoid the evils of rivalry and competition, these were, early in 1602, united into one great company—the Netherlands East India Company. An assembly of seventeen directors was charged

---

* Theal's South Africa—"The Story of the Nations" Series—Vol. 38. (London : Fisher Unwin).

† John Davis. By Clements Markham, President of the Royal Geographical Society. The World's Great Explorers Series. (London : G. Philip & Son).

with the management of this powerful corporation, which was destined not only to break the power of Portugal and Spain in the Eastern seas, but also to bring South Africa within the pale of civilisation. The Dutch were alive to the advantages of the Cape as a "Half-way House" to the East, but a proposal from the directors of the English East India Company to the "assembly of seventeen" to build jointly a fort, and form a place of refreshment on the South African coast, was not entertained, although both companies ordered the commanders of their outward-bound fleets to examine and report upon suitable sites for the purpose. The English captains decided that Table Bay was the best place, planted the English flag on the Lion's Rump, and proclaimed English sovereignty over the adjoining country in the name of His Majesty King James II. Possession, however, was not maintained, and ultimately the favourable reports of the officers of a Dutch ship, which had been wrecked in Table Bay, decided the Dutch Company to establish a victualling station for their fleets in Table Valley. Plans were drawn up and approved, and three vessels were got ready to convey men and materials to the Cape, and placed under the command of Jan Anthony Van Riebeek, who had been appointed governor of the new settlement. These vessels—the *Dromedaris*, an old Indiaman, the *Reijger*, a smaller vessel, and the yacht *Goede Hoop*, anchored in Table Bay on the 6th of April, 1652, after an unusually quick passage of 104 days from the Texel.

Mr. Van Riebeek was an irascible little man of undaunted spirit and indomitable perseverance, and for ten years he ruled the new settlement with Spartan simplicity and severity. He immediately set about building an earthenwork fort as a stronghold against the savages, and, under the protection of its guns, the settlers, 116 in number, all of whom were employés of the Netherlands East India Company, built their huts, and laid out their gardens and pasture grounds. Van Riebeek himself actively engaged in the work of cultivating the ground, and the breeding of cattle and sheep. Vegetables were raised, and wheat, barley, oats, and maize successfully grown. The vine, the orange, the olive, the mulberry, the fig, peach, apple, and other fruit trees were introduced, and young oaks and firs were brought from Europe. Cattle and sheep were obtained from the Hottentots, horses were imported from Java, and pigs, sheep, dogs, rabbits, and poultry from Europe. The settlement throve apace, and Van Riebeek was soon able to furnish the numerous vessels that called at Table Bay with abundant supplies of provisions. But everything was done by and through the Company, which had a monopoly of the trade, internal and external. No competition or free immigration was allowed at first, but the cost of so exclusive a system induced the directors to permit a few burgher families to settle, and cultivate small plots of land in the neighbourhood of the fort. These farmers, or "boers," were *the first true colonists of South Africa*. Ever-increasing numbers of other discharged servants of the Company, and immigrants from Holland and Germany, also settled on the land, and gradually extended the limits of the colony. Negro slaves were introduced

in 1658, and Asiatics, chiefly natives of Malacca, Java, and the Spice Islands, were brought into the settlement. For a few years, the intercourse between the Dutch settlers and the aboriginal inhabitants was friendly in the extreme, and when the commander or other officers of the garrison visited any of the Hottentot kraals, they were received with effusive welcome. Van Riebeek did not much relish their amicable embraces, for in his journal he says:—"We had again a suit of clothes destroyed from the greasiness of the oil and filth with which they, and particularly the greatest among them, had so besmeared themselves, that they shone like looking-glasses in the sun, the fat trickling down from their heads and along their whole bodies, which appeared to be their greatest mark of distinction."

When the "Caepmans" saw the white men ploughing their ground and taking possession of their pastures, they became alarmed, then angry; hostilities broke out; a white herdsman was killed, and several natives were shot. After months of unrest, peace was concluded, and the first of the long series of wars with the natives, which darken the story of South Africa, came to an end. On the anniversary of the founding of the settlement eight years before, the Captain and chief of the tribe, with the principal men, met Van Riebeek and his Council at the fort to discuss the terms of peace, and the arguments advanced by the aborigines were certainly unanswerable. The Dutch Commander reported that the Hottentot leaders "dwelt long upon our taking every day for our own use more of the land which had belonged to them from all ages, and on which they were accustomed to depasture their cattle. They also asked whether, if they were to come into Holland, they would be permitted to act in a similar manner, saying, 'What would it signify if you remained here at the Fort? but you come quite into the interior, selecting the best for yourselves, and never once asking whether we like it, or whether it will put us to any inconvenience. Who,' said they, 'should be required to give way, the natural owners or the foreign invaders?' They insisted much upon their natural right of property, etc., and that they should at least be at liberty to gather their winter food—the bitter almonds and roots which grew there naturally . . . and they insisted so much on this point that our word must out at last: That they had now lost that land in war, and therefore could only expect to be henceforth entirely deprived of it; that their country had thus fallen to our lot, being justly won by the sword in defensive warfare, and that it was our intention to retain it." Truly a typical example of European dealings with the natives all the world over! Van Riebeek concludes the entry with the naïve remark, that after the terms of peace were settled, the chief and all the principal people received presents of brass, beads, and tobacco, and "were so well entertained with food and brandy that they were all well fuddled, and if we had chosen we could have easily kept them in our power, but for many weighty reasons this was not deemed expedient, as we can do that at any time, and meanwhile their dispositions can be still further sounded."

The hopes of a peaceful occupation of the Cape were

thus rudely dispelled ; but the directors in Holland—to their honour be it said—issued stringent orders that the natives were to be justly and kindly treated, and their property respected. They wrote, "The discontent shown by these people, in consequence of our appropriating to ourselves—and to their exclusion—the land which they have used for their cattle from time immemorial, is neither surprising nor groundless, and we therefore should be glad to see that we could purchase it from them, or otherwise satisfy them." This was done in 1672 ; the "lands, rivers, creeks, forests, and pastures inclusive," from the Cape Peninsula to Saldanha Bay, were purchased (?) from two petty Hottentot potentates for brandy, tobacco, beads, and merchandise of the nominal value of £1,600, but actually, according to the accounts furnished to the directors, the articles transferred cost £9 12s. 9d. ! *

After this agreeable transaction, the precursor of many such transfers of real estate in South Africa, the Company posed as absolute owners of the soil, and though the Assembly of seventeen showed some lingering regard for the rights of the original owners, Judge Watermeyer tells us that scarcely ten years had elapsed, before the Dutch authorities had ceased all affectation of a desire that native claims should be respected. In 1673, war broke out between one of the most powerful of the Hottentot tribes, near the Cape Peninsula, and the settlers, and although the latter were aided by some friendly Hottentots, the settlement was practically blockaded on the land side, and the cattle trade entirely stopped ; a fact which had most important consequences for the future of the country, as it forced the authorities to encourage the breeding and rearing of cattle by Europeans, instead of being entirely dependent on precarious supplies from the natives. Simon van der Stell, one of the most famous of the early Dutch governors, threw himself heart and soul into the new project, and induced many burghers to leave the settlement by an offer of as much land as they could cultivate at Stellenbosch and Drakenstein, with extensive grazing rights. Land to farm was also granted to the government officers, and the energetic Van der Stell himself laid out and planted a beautiful wine-farm at Constantia, near Wijnberg, to which he retired in 1699, being succeeded as Governor by his eldest son, Wilhelm Adrian van der Stell, who also, together with his brother, took to farming on a large scale for his own benefit—a proceeding which subjected the free burghers to ruinous rivalry. They therefore sent a memorial to the directors in Holland, complaining of "the unrighteous and haughty tyranny" of the Governor, who had in various ways used his position to fill his own purse and those of his relatives and friends, and would only listen "to reasons that jingle." Van der Stell instantly took severe measures against the memorialists : some he banished, others he committed to prison, while some escaped and remained in hiding until the directors of the Company—"for the quieting of disorder and the restoration of tranquillity"— dismissed the Governor and confiscated his estate, and forbade their servants to "own or lease land in the Colony,

* Theal.

or to trade directly or indirectly in corn, wine, or cattle. The burghers were expressly admitted to have the same rights, as if they were living in the Netherlands"—previously they had been little better than slaves to the Company, whose officers had acted as lords and masters of the settlement in every way, and had monopolised the trade and restricted the industry of the inhabitants, levying heavy taxes on all their produce, which could only be sold to the Company at the Company's price, while at the same time they could buy nothing but what came from the Company's store. Small wonder, then, that the rule of the Company had become so obnoxious to the boers that "many of them moved away with their waggons and flocks and herds far inland beyond its control." Thus was originated that peculiar habit of *trekking*, or moving from place to place, which has always characterised the Dutch farmers of South Africa, and which led them to people Natal and to found the Overberg Republics. "At first the Government tried by threats of severe punishment to stop the migration from the seaboard, but the movement was too strong to be checked. The farmers continued to move inland, enticed not only by the thought of fresh pastures for their cattle and game for their guns, but by a desire to be free from the irksome restraints of Government. The Company made some attempt to follow the migratory colonists. A magistracy was established at Swellendam in 1745, and at Graaf Reinet in 1786, and in 1788 the Great Fish River was declared to be the boundary of the settlement.*

There have never, says Theal,[†] been people less willing to submit silently to grievances, real or imaginary, than the colonists of South Africa, and no doubt much of the impatience of restraint and love of individual liberty shown by the trekboeren, as well as by the burghers, was derived from the many Protestant Frenchmen who were sent out to the Cape in 1687. These Huguenots— "exiles for conscience sake"—had been driven from their own country by the revocation of the Edict of Nantes by the intolerant and cruel Louis XIV., and had sought refuge and received sympathy and kindness in the Netherlands, then the stronghold of liberty in Europe. The greater part of the refugees were located in the valley of the Berg River ; others were scattered over the country between the Groenberg, the Koeberg, and Hottentots-Holland. They soon had "cornfields green and sunny vines," and endeavoured to preserve their language and form of worship. This, however, was discouraged, the Company desiring that French should, in time, entirely die out, and that nothing but Dutch should be taught to the young to read and write. The use of French in communications to the Government was forbidden in 1709, and, in 1724, "the reading of the lessons at the church service in the French language took place for the last time. In little more than half a century after the arrival of the Huguenots, French had almost ceased to be spoken

* See further Russell's admirable and graphic *résumé* of South African history in—*Natal : The Land and its Story.* (Pietermaritzburg : P. Davis and Sons. London : Simpkin Marshall and Co.).

† Theal's *South African.* "Story of the Nations" Series. (London : Fisher Unwin).

among their children, they, by marriage and social connections, had become merged with the Dutch and Germans around them, using the Dutch language only."*

The descendants of these Huguenots are widely scattered all over South Africa, and Huguenot names such as De Villiers, Du Toit, Jourdan, Retief, Theron, Joubert, etc., are common to this day from the Cape to the Zambesi. Many of the wine farms and estates also bear French names—the Huguenots excelled as wine growers, and gave a great impetus to the cultivation of the vine, although they were not, as is commonly supposed, the founders of viticulture at the Cape.

After the formal purchase of the Cape peninsula and adjoining district from the Hottentot chiefs—who probably thought they might get something for what would otherwise be taken for nothing—consideration for the rights of the natives dwindled away ; and as the settlers advanced, the Hottentots retired. The Namaquas and the Chocoquas went north and settled in the wild and arid country on either side of the lower Orange, an undesirable region, where their descendants still dwell. Tribal feuds, constant war with the savage Bushmen, and disease, decimated the other tribes. Small-pox, which made its first appearance in South Africa in 1713, first seized the Negro slaves, then attacked the Europeans, one-fourth of whom died ; but it proved most fatal to the Hottentots, whole clans of whom in the neighbourhood of the Cape were almost entirely swept away. The miserable remnants of these tribes struggled for existence on reserves set apart for their use, and were employed on the cattle stations and farms.

When the Dutch arrived at the Cape, a primeval hunting people, armed with bows and poisoned arrows, was thinly scattered over all the wilder parts of the country to the south of the Orange, and between these wild Bushmen and the Hottentots there had always been a fierce and deadly feud. The Bushmen, who were armed with bows and poisoned arrows, were still more enraged at the invasion of their happy hunting-grounds by the white men, and revenged themselves by making frequent raids on the farms, killing the herdsmen and stealing the sheep and the cattle, and occasionally looting the homesteads and murdering the inmates. As the farmers advanced, the Bushmen retired and sought refuge in the mountain fastnesses along the northern border, and for thirty years the sorely-harassed farmers carried on what was practically a war of extermination. The records of Graaf Reinct show that, between 1786 and 1794, more than 200 persons were murdered by the Bushmen, while the "commandos," or armed bands of farmers, who, aided by the Hottentots and half-breeds, scoured the country along the great mountain range, had shot over 2,500 Bushmen. These wild pigmies never would surrender, and they were as fiercely hostile to the white man as to the Hottentots and Kaffirs. The Bushman was an Ishmaelite indeed ; his hand was against every man, and every man's hand against him. Despised and yet dreaded, these untamable pariahs of South Africa were hunted from kloof to kloof, and shot down with as little consider-

ation as if they had been wild animals. Few of them are now left on the northern border of the Cape ; and from the Drakensberg and the Maluti Mountains, where, fifty years ago, they were comparatively numerous, they have entirely disappeared. Numbers of them still roam over the desolate plains of the Northern Kalahari, and Selous tells us that they are unrivalled as assistants and trackers in the hunting veldt.

The Gamtoos River, which had formed the eastern limit of the Hottentot region, was early adopted by the Dutch as their frontier towards the east, and many of the more adventurous Boers crossed the river, and thus came into conflict with the Kaffirs, a people who soon proved to be much more formidable neighbours than the pigmy Bushmen or the degraded Hottentots. The more advanced of the Kaffir clans—the Kosas—had, in 1779, crossed the Fish River, but did not at first molest the Europeans, who had settled in what was then beyond the limits of the colony, although they murdered a number of Hottentots and took their cattle. Becoming bolder, the Kosas began to drive off the cattle of the Boers also, upon which they were attacked and dispersed. Again the Kaffirs invaded the colony, and this time in such numbers that a commando of the border farmers was called out and placed under the command of Adrian van Jaarsveld, who was ordered to drive back the Kaffirs across the Fish River. With a small force of 92 burghers and 40 Hottentots, all mounted and well armed, Van Jaarsveld fell upon the Kaffirs and smote them hip and thigh. In less than two months not a single Kosa was to be found to the west of the Fish River, and the *first Kaffir war* was over.

For nearly a century the Dutch East India Company had been supreme in South Africa, but during the administration of Van Tulbagh (1750-1771), although the colony was becoming more and more prosperous, the power of the Company began to decline, and ultimately the heavy losses sustained by the casting away of many of its richly-laden merchantmen in Table Bay, where in the winter season they were exposed to the full fury of the north-west gales, together with the mismanagement of affairs in other of its possessions besides the Cape, and, above all, the growing competition of the English and French in the markets of the East, brought the once powerful and rich Company into a state of hopeless insolvency. In South Africa, the arbitrary rule of Tulbagh's successor, Van Plettenberg (1771-1785), and the corruption and exactions of the Company's subordinate officials, caused much disaffection among the colonists, who sent delegates to Holland to obtain redress, only to be told by the Directors that the settlers had been permitted "as a matter of grace to have a residence in the land and to gain a livelihood as tillers of the soil, and that the settlement was planted not for their commercial advantage, but for the welfare of the Company." The Company, in view of the unstable state of affairs, decided to station a large body of troops at the Cape and to fortify the peninsula, so that it should be secure from invasion. Van Plettenberg was therefore recalled, and an engineer officer—Colonel Van de Graaf—appointed governor. His reckless expenditure, which necessitated increased taxation and a forced paper currency, petty acts

---

* Noble.

of tyranny, constant troubles with the Kaffirs on the eastern frontier, deepened the dissatisfaction of the colonists, and caused an agitation against the Company which did not cease until the despotic rule of this famous corporation came to an end.

During the struggle between England and her American colonies, Holland had joined France and the armed neutrality powers, and in 1780 the British Government declared war against Holland, and planned an expedition to seize the Cape. This became known, and a French fleet was immediately sent out with troops to aid in its defence. Off the Cape Verde Islands, the out-going English fleet was accidentally met, and was fiercely attacked and partially disabled. The French commander then made all sail for the Cape, and landed his troops a month before the English fleet arrived. So complete were the arrangements for the defence of Cape Town, that the English admiral did not venture to attack the place ; but contented himself with seizing the richly-laden Dutch Indiamen in Saldanha Bay. This brief war ended in 1783, but, before the conclusion of peace, the Dutch mercantile marine was almost annihilated.

In 1786, De Graaf founded the town of Graaf Reinet, and formed a new district between the Gamtoos and the Fish rivers, and proclaimed the latter stream as the eastern boundary of the colony. This forward move was speedily followed by a renewal of the troubles with the natives, who suddenly crossed the Fish River in March, 1789, and began to drive off the farmers' cattle. A commando was called out, but was disbanded without firing a shot, by order of the Government. In a short time the Kosas recommenced their raids—seizing the cattle, burning the homesteads, and murdering every farmer they came across. The burghers were again assembled, but although the Kaffirs had laid waste the coastlands as far as the Zwartkops River, and had driven off 65,000 head of cattle, the campaign was an utter failure, owing to the action of Maynier, the landdrost of Graaf Reinet, who evidently believed that smooth words would turn away wrath and—Kaffirs. And so, much to the chagrin of the farmers, the *second Kaffir war* ended in a delusive peace.

This method of dealing with the Kosa marauders added to the prevailing discontent, and the mismanagement of affairs generally made the Company's government so obnoxious, that great numbers of Boers trekked away with their waggons and flocks and herds far inland beyond its control, while the farmers on the Eastern border and in the valley of the Breede broke out in actual rebellion. In February, 1795, the burghers of Bruintjes Hoogte assembled at Graaf Reinet, and declared themselves independent. They expelled the landdrost, and set up a republic of their own, with Adrian van Jaarsveld as military commander. The Commissioner-General, Sluysken, who had been put in charge of the colony when the Directors recalled the spendthrift Van de Graaf, could do nothing to stop the movement ; and in June, the burghers of Swellendam also expelled their landdrost, and elected a "national assembly." Sluysken had no force to send against these tiny republics, his treasury was empty, and the people elsewhere were mutinous ; besides, he had the

native question to deal with, and a probable invasion to prepare for.

The Cape was really in a state of anarchy, when an unexpected event happened. France had passed through the throes of a revolution, which had deluged that fair land with blood and with crimes. In Holland, a strong "patriot party" was in sympathy with the French revolutionists ; and, in 1795, the armies of the Convention overran the country, and forced the Stadtholder, the Prince of Orange, to fly for refuge to England, where he continued to co-operate with the Allied Powers against France. It was feared that the Cape, the most important of all the maritime stations on the ocean-route to India and the East, might fall into the hands of the French. The British Government, therefore, decided to occupy the country, and a mandate was obtained from the fugitive Stadtholder commanding the authorities at Cape Town to receive the English fleet, and to admit the English troops into the Castle and the forts. Sluysken had meanwhile received orders from the Company's directors to oppose the landing of any force, British or French ; and when the English fleet under Admiral Elphinstone, with a strong body of troops under General Craig, arrived in Table Bay, he refused to admit them. After a very feeble defence, however, he surrendered the castle and town ; and thus, on the 16th of September, 1795, the detested rule of the Dutch East India Company in South Africa, after an occupation of one hundred and forty-three years, came to an inglorious end.

For the next eight years the Cape remained under British military rule, but although all monopolies and restrictions on trade, from which the colonists had so grievously suffered during the Dutch occupation, were removed, and large sums of money were freely spent in the Colony by the British Government, the country people generally were in a state of chronic rebellion. General Craig, who had assumed the government, did everything in his power to soothe the susceptibilities of the sturdy colonists. Obnoxious taxes were repealed ; the paper money was taken up at its full nominal value ; and the farmers were told that they could now buy and sell freely, and that any complaints they had to make would be attended to. No opposition was offered to the new government in the Cape and Stellenbosch districts ; and the people of Swellendam abolished their republic, but the burghers of Graaf Reinet did not submit to the English until their supplies of ammunition and goods were cut off. In the meantime, the States-General had sent out a squadron of nine vessels, with 2,000 troops on board, to aid the colonists against the English. The Dutch Admiral put into Saldanha Bay, where he was caught, as in a trap, between a strong British fleet and a large British army. He surrendered without even an attempt at resistance.

The conciliatory attitude of General Craig was unfortunately not shown by his successors ; and the strict rule of the Earl of Macartney, who forced the burghers to take the oath of allegiance or leave the country ; the arrest of the old commander Van Jaarsveld, and his rescue by the farmers of Graaf Reinet ; the "friendly arrangements" which ended the *third Kaffir war*,

although the whole country from the Sunday's River westward even to Langekloop and the Knysna had been desolated by the Kaffirs and their Hottentot allies ; the thoroughly corrupt administration of Sir George Yonge, a man who could only be approached through unscrupulous favourites—all combined to embitter the Dutch against the British, and to make welcome the change when, in February, 1803, the colony, according to the terms of the Treaty of Amiens, was restored to its original owners. The population of the colony at this time is supposed to have been about 70,000, of whom 22,000 were Europeans, 26,000 slaves, and the rest Hottentots.

No commercial company now intervened between the colonists and their Fatherland ; and the new Governor, General Janssens, and the Dutch High Commissioner De Mist, proved to be liberal and noble-minded men, who did their utmost to increase the prosperity of the colony and to elevate every class of the population. The farmers were encouraged to plant trees and preserve forests, to establish schools, to treat the Hottentots as a free people and the slaves as fellow-creatures, and to give the Kosas no cause for making raids.

Events in Europe, however, prevented the full fruition of these wise and humane measures. In less than three months after the restoration of the colony, the Batavian Republic and France were again at war with Great Britain, and General Janssens devoted all his attention to the defence of the Cape Peninsula against any attempt that might be made to recapture the colony. In January, 1806, an English fleet anchored at the entrance to Table Bay, and an army of 7,000 men, under General Sir David Baird, landed on the Blueberg beach. General Janssens, with a motley force of burghers, Dutch soldiers, German mercenaries, French seamen, Malays, Hottentots, and even slaves, endeavoured to bar the way to Cape Town. but the flight of the German mercenaries forced him to retreat. Two days later, General Baird entered Cape Town, and, on the 18th, General Janssens capitulated on honourable conditions ; and thus the Cape again became a British possession, though it was not until 1815 that the King of the Netherlands, in consideration of a payment of six millions sterling, finally ceded the Cape, along with the Dutch settlements in Guiana, in perpetuity to the British crown.

There were in the Colony at this time about 30,000 slaves, chiefly the descendants of those introduced in the early days of the settlement from the Guinea Coast. One of the first acts of the new Governor, the Earl of Caledon, was to abolish the slave-trade, and at the same time to alleviate the condition of the Hottentots and other coloured people. Absolute democrats and lovers of freedom the Boers might be, but their ideas of liberty, equality, and fraternity did not include the Kaffir and the Hottentot. All the coloured races were "black property," or "creatures :" and anything approaching equality between the Boer and the black was, and still is, an impossible idea. "The more ignorant of them believed themselves to be 'God's chosen people,' and the Bushmen and the Kaffirs the 'Canaanites,' whom they had a divine command to smite and utterly destroy.* The cruellest raids on the

natives in the interior by the Trekboeren in after years were publicly justified by the instructions given to the old Jewish warriors.* The Boers were convinced that in conquering, dispossessing, and enslaving the natives, they were obeying the behests of the Almighty."†

The Governor also opened up postal communication with the inland districts, and established Circuit Courts, before which, in 1812, numbers of Boers were accused by the missionaries of ill-treating the natives, and some were convicted, with the result that a very bitter feeling was aroused among the farmers, "it being an unheard of thing that a European should be punished for an assault on a native." The Earl of Caledon's successor, Sir John Cradock, also caused much animosity by his proclamation recommending the study of the English language.

In 1811-12 occurred the *fourth Kaffir war*, which was occasioned by an irruption of some Kaffir clans into the "neutral territory" between the Sunday's River and the Fish River. Colonel Graham, who was in command of the British and Colonial forces, sent an officer to try and persuade the Kosas to retire peaceably, but the fiery old chief Ndlambe angrily retorted, "This country is mine ; I won it in war, and intend to keep it." The order was then given to advance, but before an attack was made, Landdrost Stockenstroom and eight farmers were treacherously murdered during a conference with a number of Kaffir warriors. This led to terrible retaliation. No mercy was shown to any of the warriors who resisted, no prisoners were taken ; it was war to the knife until every Kaffir had been driven across the Fish River, along which a line of forts was erected, the principal post being named Graham's Town in honour of the officer in command.

A most important concession in favour of the farmers was made by proclamation in 1812. All holders of *lands on lease* were allowed to convert them into perpetual quit-rent properties ; in other words, farmers, who had hitherto held their land on yearly lease from the Government, were made absolute owners of their farms. But this and other boons and blessings bestowed by the British Government failed to conciliate the Boers, and, a year after Lord Charles Somerset succeeded to the governorship, the more turbulent of them broke out in open rebellion, which resulted in five of them being hung at Slagter's Nek. The horrible circumstances which attended the execution of these unfortunate men deepened the animosity of their fellow-countrymen against British rule.

In spite of forts and troops and burgher commandos, the feeling of insecurity along the eastern frontier was so general, that offers of free farms even of 4,000 acres failed to attract many settlers. Fleet-footed Kaffirs again and again slipped over the border during the night, and "lifted" many a fine herd of cattle, the soldiers retaliating by an occasional raid on the nearest kraal, and seizing indiscriminately all the cattle within reach. This game of hide and seek increased the hostility on both sides, and led to the *fifth Kaffir war*. In 1817, the

---

* Russell's *Natal : The Land and its Story.*

* Deut. xx. 10-14, and similar passages.

† See further "Livingstone and the Exploration of Central Africa," by H. H. Johnston, C.B., H.M. Commissioner for Nyasaland.—The World's Great Explorers Series. (London : George Philip & Son).

Governor had recognised Gaika as the supreme chief of the Kaffirs to the west of the Kei River, hoping through him to control the other chiefs. They, however, headed by Ndlambe and a famous seer named Makana, or "Lynx," refused to acknowledge Gaika as over-lord, whereupon a desperate battle was fought on the Debe flats, Gaika's forces being driven from the field with frightful slaughter. The fugitive chief appealed to the Government for aid. A force of burghers and soldiers went to his assistance. Ndlambe's kraals were destroyed, and his cattle driven off, but the dense thickets sheltered his warriors, who immediately the troops retired, fell again upon Gaika, and then poured into the Colony, murdering the whites and Hottentots, and destroying their property. Led by Makana, a man of conspicuous ability and daring, who aimed at uniting all the western clans into one strong nation, ten thousand savage warriors swooped down upon Graham's Town, in three columns. Their impetuous rush was stopped by a deadly fire of artillery and musketry, and, after a brief struggle, the discomfited warriors were driven back with heavy loss. Two thousand soldiers and burghers followed them into Kaffirland, and hunted them out of the bushy fastnesses of the Chumie and the Keiskamma. Makana, with the magnanimity of a Roman, voluntarily surrendered himself. Walking calmly into the British camp, he said, " If I have occasioned the war, let me see if delivering myself to the conquerors will restore peace to my country !" Gaika was restored to his lands, and, in 1820, he ceded the country between the Fish River and the Keiskamma to the Colony.

Lord Charles Somerset was very anxious to see the vacant lands in the eastern frontier districts—which he described as "unrivalled in the world in beauty and fertility "—occupied ; and, on his recommendation, the Imperial Parliament voted £50,000 towards their colonisation. In a very short time over 90,000 people applied for passages, 5,000 of whom were accepted and sent to South Africa. In April, 1820, the *Nautilus, Ocean*, and *Chapman* arrived in Algoa Bay with the first batch of immigrants, and were followed by 23 other transports. The landing place, then a small fishing village, was named Port Elizabeth, after the acting Governor's wife ; and thence the settlers were distributed over the pleasant country between the Bushman's and Fish rivers and the Zuurberg and the sea. In spite of much distress and inevitable difficulties, the settlers in a few years became prosperous, and Port Elizabeth, the chief port, and Graham's Town, the chief inland centre of the district, grew from mere hamlets into populous and flourishing towns.

A few years later, a series of sweeping changes irritated the old Dutch colonists almost past endurance. In 1827, English was ordered to be used instead of Dutch in all official proceedings and business ; and in the following year, the courts of justice were remodelled after the English pattern, the Burgher senate was abolished, and English resident magistrates and civil commissioners took the place of the landdrost and heemraden, who had hitherto administered justice and managed local affairs in the country districts. Everything was becoming so English, that the Dutch began to feel as if they were no longer in their own country. Dutch ideas with regard to the natives also received a severe shock when, in 1829, by an Order in Council, it was enacted that "all Hottentots and other free persons of colour lawfully residing within the colony are in the most full and ample manner entitled to all and every right, benefit, and privilege to which any other British subjects are entitled." This "Magna Charta" of the natives was followed by the emancipation of the slaves, which was carried into effect by Sir Benjamin D'Urban in 1834. The 35,000 slaves in the Colony were officially valued at three millions sterling, but the Home Government only allowed a million and a quarter as compensation to the owners, and a large part of this sum never reached the hands of the indignant Boers, many of whom refused to receive any of it. The proverbial last straw was the reversal by the Earl of Glenelg, who became Secretary of State for the Colonies in 1835, of what the colonists considered to be the only safe policy of dealing with the Kaffirs, after a war which forms one of the saddest chapters in the troubled history of the Cape.

Space will not permit us to trace the events which led to the *sixth Kaffir war*, which broke out in December, 1834. On the 22nd of that month, a horde of between ten and twenty thousand savages under Hintsa suddenly rushed over the border, and spread terror and destruction over the whole country. In a week, fifty farmers were murdered, 450 farmhouses burned, and 4,000 horses, 100,000 head of cattle, and 150,000 sheep were driven off. Most of the British settlers of 1820 were reduced to destitution, and many of them, failing to reach any place of refuge, were barbarously murdered. The consternation in Cape Town and throughout the colony was intense. Every available soldier and burgher were hurried to the front, and Colonel (afterwards Sir Harry) Smith pushed the war into the enemy's country, and, after severe fighting, succeeded in forcing Hintza to sue for peace. He was shot dead soon after, while endeavouring to escape, and was succeeded as paramount chief of the Kosas by his son Kreli, with whom peace was concluded. British authority was then proclaimed over the territory of the conquered clans as far as the Great Kei, while the Governor brought some 18,000 Fingoes—remnants of the Fetcani, or Zulu refugees, who had been enslaved by the Gcaleka Kaffirs—out of Kaffirland and located them between the Keiskamma and Fish Rivers, so as to form a "buffer" between the Kosas, who hated them bitterly, and the colonists, upon whom they depended for protection. The Western Kosas were now British subjects, under the control of Colonel Smith, who, with the troops, was stationed at a place that grew in later years into the important town of King William's Town.

These arrangements promised to work well, but there was a clique in Cape Town that disapproved of the Governor's plans. Its chief, Dr. Philip, the champion of the natives against what was stigmatised as oppression and cruelty, visited England, and impressed his views on the Secretary of State, who wrote to Sir Benjamin D'Urban to the effect that the frontier must be retroceded and western Kosaland given up ; and that he considered the Kaffirs, as the victims of systematic injustice

through a long course of years, were amply justified in rushing into war, and had a perfect right to endeavour to extort by force that redress which they could not expect otherwise to obtain. The Governor stoutly protested that the reversal of his policy could not but be pregnant with insecurity, disorder, and danger ; but his expostulations only procured his dismissal. The British settlers, who had suffered so much, also protested, but in vain. As for the aggrieved Dutch colonists, disdaining to make a vain protest and unable to offer any effective opposition, they determined to quit the country and to seek somewhere in the wilderness beyond the Orange and the Vaal—devastated and unpeopled by the impis of the merciless Tshaka—a new home beyond the control of the hated British.

The wars and devastations of Tshaka, the founder of the dreaded Zulu power, form a terrible answer to the thoughtless utterances of those who echo the parrot cry of "Africa for the Africans." In spite of all actual and seeming harshness and cruelty on the part of both Boers and Britons in South Africa, no fair-minded enquirer but will admit that the advent of the white man in South Africa has been the salvation of the black. Bantu potentates, of the type of Tshaka, Dingaan, Moselekatse, Cetywayo, and Lo Bengula, have slaughtered millions of their fellow-countrymen, have made many a populous and flourishing region a desolate waste, and but for the white man and his dreaded rifle, millions more would have been similarly exterminated, and the horrid slaughter would, in all probability, have continued until the southern extremity of the continent had been all but depleted of human, as it has been of animal, life.

Tshaka, a military genius of the highest order but a sanguinary despot, had raised the insignificant Zulu clan into a powerful nation. This sable Napoleon, like his European counterpart, aimed at universal sovereignty, and, by a series of ruthless aggressions and wholesale massacres, had become paramount chief of all southeastern Africa, from Kaffraria to the Limpopo. During his reign probably a million people were slaughtered by his savage impis. "He turned thousands of square miles into literally a howling wilderness, shed rivers of blood, annihilated whole communities, converting the members of others into cannibals, and causing misery and suffering, the full extent of which can never be known."

When Tshaka commenced his reign of terror, Natal was a black Arcadia, inhabited by no less than 94 tribes, representing about a million of people, living in peace and plenty ; but in a few years the once "incredibly populous" land was a desolate wilderness. In the dense thickets and mountain gorges were hidden a few thousand miserable starvelings, subsisting on wild fruits and roots—some even on human flesh. In 1824, three bold Englishmen—Lieutenants Farewell and King, and Mr. Henry Fynn—settled on the shores of the Bay, and, perhaps by their very boldness, won the favour of the ferocious Zulu despot, who indeed went so far as to cede to them in perpetuity a tract of land along the coast, including the Bay, and the country inland to the Drakensberg. Refugees from Tshaka's tyranny flocked to the

English settlement, and communications were opened with the Cape.

One of Tshaka's ablest generals, a chieftain named Moselekatse, had also to fly from his master's vengeance. Having, Ananias like, kept back part of the booty he had taken in a successful raid, the enraged king sent an army to put him and his soldiers to death. Being warned in time, Moselekatse and his followers fled over the Berg, and began to devastate the upland plains on both sides of the Vaal, ultimately setting up his military kraals in the valley of the Marikwa.

Thousands of the wretched, timorous, and unwarlike Bechuana tribes were slaughtered in mere wantonness by the fierce Matabele, as the Zulu hordes of Moselekatse were named by the wretched tribesmen, of whom the Batlapins of Kuruman alone escaped, saved by the presence among them of the devoted missionary Moffat, who visited Moselekatse, and won his respect.

Remnants of many broken tribes had, in the meantime, gathered round an able and astute young chief named Moshesh, in the mountainous enclave now called Basutoland. Moshesh fixed upon an impregnable mountain stronghold—Thaba Bosigo—as his capital, and thence consolidated and built up a formidable native power. In 1831, a Matebele army besieged the Basuto stronghold, but had to retreat. Moshesh magnanimously sent his thwarted foes a present of fat oxen, with a complimentary message, which so astonished the Matabeles that they never again attacked the Basutos. This was in 1831, in the same year that Dingaan, who with one of his brothers had murdered Tshaka and had assumed the chieftainship of the Zulu nation, declared Henry Fynn, the survivor of the three English pioneers, the "Great Chief of the Natal Kaffirs." Dingaan had not the military genius of his brother Tshaka, but he was equally bloodthirsty, and even craftier and more treacherous.

The *Great Trek* of the discontented Boers commenced in 1835, and in two or three years not less than ten thousand people left the Cape Colony with their waggons and oxen, their horses and cattle and sheep and goats, and in detached parties, of from fifty to one hundred families, trekked slowly across the wide plains of the Orange and the Vaal. Their leaders were "grave, stern men, imbued with the spirit of the Dutch burghers who defeated Alva, and of the Huguenots who fought under Condé. The Bible was their only literature. No important undertaking was ever entered upon without prayer and praise being offered to the Almighty. Like the Puritans, they had as much faith in the psalm as in the pikepoint."[*]

The plains were then covered with myriads of antelopes and quaggas, and over two hundred lions were shot during the trek. The Basuto chief, Moshesh, offered no opposition to the progress of the emigrants ; but Moselekatse's fierce warriors fell upon two small detached parties when near the present town of Kroonstad, massacred nearly all of them, and carried off their waggons and flocks and herds. Succeeding parties were more cautious, and never encamped without drawing their waggons close together in a square or circle, filling up the openings between and

* Russell.

under the waggons with thorn bushes. When attacked, men and women, and even children, fought with desperate energy, and, in spite of the overwhelming numbers of their savage foes, succeeded in driving them off. To avenge the murder of their comrades, one hundred farmers rode across the Vaal and attacked Moselekatse's headquarters at Mosega, inflicting upon him so severe a defeat that he fled with his warriors to the highlands between the Limpopo and the Zambesi. From this new Matabeleland, impi after impi continued to raid and slaughter surrounding tribes in the old Zulu style, until the advent of the Chartered Company and the conquest of Lo Bengula, the son and successor of Moselekatse. Another of Tshaka's generals, rather than return home to certain death, after failing to carry out an order to drive the Portuguese from Delagoa Bay, went north with his warriors across the Limpopo and occupied the country now called Gazaland, which is still under the undisturbed rule of his successor, Gungunhana.

From the upland plains, an advanced party of the emigrant farmers, under Pieter Retief, made their way down the wild passes of the Drakensberg, through a seemingly unoccupied country, to the Bay, where they were warmly welcomed by the English traders. Retief and a few men rode across the Tugela to Dingaan's kraal, and asked the Zulu king for a grant of land in Natal. Dingaan agreed, if the Dutchmen would only prove their goodwill by recovering some cattle, which a Rob Roy of the Berg had carried off. This was done, and in high glee Retief, accompanied by about 70 horsemen and 30 servants, returned to the royal kraal of Umgungundhlovu. There they were received in the friendliest manner ; a formal deed of cession was drawn up and signed, and the Dutchmen were getting ready to saddle-up and leave, when Dingaan invited them to drink *utywala* with him in his great place. Suspecting nothing, the farmers complied, and even left their muskets outside the enclosure ; but, while seated on the ground, they were suddenly seized by order of the treacherous savage, and dragged to the place of execution, and there done to death with knobkerries.

In the meantime, large numbers of Boers and their families had come down the Berg, and had encamped here and there over the uplands along the Tugela and Bushman's rivers. No danger was apprehended, no laagers were formed ; but in the dead of night, when all were asleep, almost simultaneously the encampments were rushed by armed Zulus, who indiscriminately butchered men, women, and children. A few escaped and warned neighbouring parties, who, hastily forming waggon-laagers, were able to beat back the masses of savages. Rendered desperate by the sight of their mangled kinsfolk, and burning with revenge, the farmers charged the Zulus, and put them to utter rout. Hundreds of the savages were struck by the avenging bullets of the farmers during that terrible flight down the Bushman's River valley. In a week, six hundred men, women, and children had been massacred. But they were indeed avenged when, on "Dingaan's Day," the 16th of December, 1838, the Boers under Pretorius killed three thousand of the warriors who attacked their laager on the "Blood"

river, and thus broke the power of the Zulu tyrant. Pushing on to Umgungundhlovu, the farmers found the royal kraal deserted and burnt ; and there, on the "hill of death," were the skeletons of their murdered friends. "Retief was recognised by his clothes, and by the leather hunting-bag slung round his shoulders. In it was found, clean and uninjured, the document by which Dingaan ceded Natal to Retief and his people for their everlasting property." Soon afterwards, the "humbled bloodhound" was again defeated by the Boers, aided by his brother Panda. Dingaan then fled to the Swazi country, where he was tortured to death, while Panda was crowned king by the victorious burghers, who now found themselves in possession of Natal, where they proceeded to lay out the town of Pietermaritzburg, and to form settlements at Durban on the coast, and Weenen up-country. A Volksraad was elected, and magistrates were also appointed. In 1840, they hoisted the flag of the "Republic of Natalia," which, however, the British Government refused to acknowledge. After serious hostilities between the English troops at the Port and the Boer forces, the Republic was abolished on the 10th of May, 1843. Two days later, Natal was proclaimed a British Colony, and in December, 1845, was annexed to the Cape. The Dutchmen of Natal thus found themselves again under British rule, and, of course, another exodus began. Some of the farmers trekked over the Berg and joined their friends in the Orange River Sovereignty ; others settled in the territory between the Vaal and the lofty Magaliesberg ; but many of them went no further than Klip River and the Biggarsberg.

The British Government, swayed by popular feeling and prejudiced by the anti-colonist action of the great missionary societies, regarded the persistent trekking of the Boers into the interior with little favour, and would fain have compelled them to return to their old homes and prevent others from leaving the colony. This, however, could not be effected by any direct means, but indirectly it was hoped that the creation of a girdle of large native states along the borders of the colony would, by cutting off communications with the emigrants, force them to return. This fatuous project of native treaty states was carried out ; and the Basuto chief Moshesh, the Griqua captain Adam Kok, and the Pondo chief Faku, were subsidised and supplied with arms and ammunition. On the same principle, the Zulu chief Panda was treated as an independent sovereign, and, like the other chiefs, was permitted to build up a power that cost much blood and treasure to cope with in later years. The sturdy farmers took little notice of these native puppets, and came and went as before. Those nominally under Adam Kok refused point-blank to acknowledge the authority of a half-bred Griqua captain. The Governor of the Cape sent a military force to aid Kok, and the farmers, taken unawares, were forced to submit, but were placed under an English officer, who fixed his residence at a place called Bloemfontein.

Meanwhile matters on the eastern frontier of the Cape were in a very critical state, and the dislike of the colony to the Glenelg policy was fully justified when, in 1846, after the country, as far west as the Sunday's River had

for ten years been harried and wasted, the *seventh Kaffir war* broke out. This war, which is known as the "War of the Axe," was brought on by the forcible rescue of a Kosa prisoner, who had been arrested for the theft of an axe and the cruel mutilation and murder of a Hottentot, to whom he was manacled. A large force started for Sandili's kraal, but the Kaffirs fell upon the waggon train, and the troops had to retreat precipitately, whereupon Kosa and Tembu warriors poured into the colony, plundering as usual. Another train from Graham's Town was captured, but, after a severe struggle, Sandili surrendered. The enormous expense of this war, and the almost insuperable difficulty of directing the government of the Cape from England, convinced even Downing Street that it would be better in every way to allow the Cape colonists to manage their own affairs.

The year following, Sir Harry Smith, who proved to be an able administrator and an impetuous commander, was sent out as Governor and High Commissioner. He at once reversed the Glenelg policy, extended the limits of the Colony on the north and the east, and formed a new province, British Kaffraria, between the Keiskamma and the Kei. Hurrying north, he put an end to the Griqua and Basuto treaty states, and proclaimed as British the whole territory between the Vaal, the Orange, and the Kathlamba Mountains, under the name of "The Orange River Sovereignty." But these wise and statesmanlike measures came too late. The Governor had scarcely returned to Cape Town when he heard that the farmers in the Sovereignty had risen in arms, with Andries Pretorius at their head, expelled the English Resident from Bloemfontein, and declared themselves independent. With characteristic energy, Sir Harry Smith hastened with all the available forces in the colony, and, meeting Pretorius and his followers at a place named Boomplaats, he defeated and dispersed them. The more violent Boers crossed the Vaal without further fighting, and Pretorius became Commandant-General of the new republic formed there, the independence of which was acknowledged by the Sand River Convention of 1850. English rule had been re-established in the Sovereignty, but the latent spirit of rebellion among the farmers, troubles with the Basutos, who more than once defeated the troops sent against them, and the general desire in England to withdraw from all interference in affairs in the interior, led the Home Government in 1854 to "abandon and renounce all dominion and sovereignty over the Orange River Territory," and to guarantee the future independence of the Orange Free State.

At this time, also, a liberal constitution was granted to the Cape. The change from an arbitrary to a representative government was most gratifying to the colonists, but they were not altogether satisfied until, in 1872, they obtained responsible government, and secured the full and free management of their own affairs.

Just before the grant of a free parliament, there occurred an event which brought the Cape into prominent notice at home. The Secretary of State proposed to make the Colony a penal station, but the people protested so strongly that, after a six months' struggle, the ship *Neptune*, which had arrived with 300 convicts in Simon's Bay, was ordered to leave, and since then no similar attempt has been made.

This violent anti-convict contest had scarcely ended, when Sir George Grey's policy of gradually increasing British control over the natives throughout Kaffirland caused Sandili and other chiefs to assume a defiant attitude. Aided by some Tembus and Hottentots, the Kosas commenced the *eighth Kaffir war*—the longest and most costly of all the native wars of the Cape—by attacking a body of troops in the Boomah Pass, and massacring a number of settlers in the military villages of the Chumie Valley. The frontier districts were ravaged ; and so fierce a guerilla warfare was kept up in the Amatolas, that it took three years' hard fighting and an expenditure of three millions sterling to suppress it. It was while conveying troops to assist in this war that the steam transport *Birkenhead* struck on a reef off Danger Point, and gave to the world that noble example of true heroism—four hundred British soldiers drawn up on deck as if on parade, and standing calmly, without a murmur, while the boats put off with the women and children and the sick people ; and then, just as the ship sank, leaping into the sea, there to perish.

The Kaffirs, however, were not really subdued ; and, in 1857, occurred the *cattle-killing mania*—a gigantic imposture instigated by the crafty Gcaleka chief Kreli, who thus hoped to throw an irresistible mass of famishing and desperate natives across the border. Moved thereto by Kreli, a witch doctor named Umhlakaza, through the medium of his niece, Nongkause, prophesied "an approaching resurrection from the dead of all the old chiefs and their followers, who would unite with the tribes to drive the white men and the Fingoes out of the country, and restore the glory of the Amakosa nation." But to this end the tribesmen must utterly destroy their cattle and their corn. This they did, and, half mad with excitement and hunger, the Kosas waited ardently for the day of resurrection ; but in vain did their eager eyes scan the horizon, none of the predicted signs appeared. Fierce fury then gave place to mad despair, and the foolish people, now perfectly destitute, died of starvation and disease in thousands ; while the strong forces that had been posted along the frontier to check the expected torrent of warriors, aided rather than checked the invasion of the colony by a continuous stream of emaciated beings, who staggered round the farmhouses, begging in piteous tones for food, which was freely given. About 30,000 Kaffirs were thus scattered over the Colony ; about 25,000 died, and large tracts of land became vacant, upon which the Governor located members of the disbanded Crimean Anglo-German Legion, who were soon afterwards joined by over 2,000 settlers from North Germany. By their industry and thrift, British Kaffraria prospered exceedingly ; King William's Town grew into an important town ; and, in 1865, the then separate colony of which it was the capital was annexed to the Cape Colony. Since then the rest of the Transkeian territories, between the Kei and Natal, have been added to the Cape Colony, but more as dependencies than as integral portions like British Kaffraria and Griqualand West. Four tribes — the Pondos, Pondomisis, Tembus, and the Kosas—occu-

pied this fine country two hundred years ago, and their descendants still own the greater part of it; but other tribes—Fingoes, Griquas, &c.—have been located there by the Government at various times, and, as may be supposed, are heartily hated by the older tribes.

For twenty years after the cattle-killing mania there was peace, but the jealousy between the Gcalekas, originally a section of the Pondo tribe, and the loyal Fingoes, brought about the *ninth Kaffir war.* Bands of Kosas swept off the Fingoes' cattle, and in February, 1878, the British camp at Kentani was charged with dense masses of warriors, who, however, were driven back. Kreli at once fled over the Bashee, and, some months later, Sandili was killed in action. Other clans rose against the Europeans in October, 1880, but were soon subdued. Pondoland, which was constantly convulsed with quarrels between rival clans, was the last portion of the Kaffir country to come under direct British authority. It was annexed to the Cape in 1894. Several islands along the coast, to the north of the Orange, also belong to the Cape, and in 1884, Walfish Bay was formally annexed to the colony. With the exception of this bay, and a little tract of land round it, the whole of the vast territory of Namaqua-Damaraland, extending along the coast from the Orange to the Cunene, and inland to the Kalahari and the upper Zambesi—a region of over 350,000 square miles in area, forms a German Protectorate. The Cape Government had long wished to take formal possession of this immense territory, but the procrastination of the Home authorities permitted Germany, on the slenderest pretext, to step in. To the great indignation of the Cape Colonists, who had long regarded the country as practically one of their own dependencies, Germany, in 1884-6, extended her claims over the whole of the vast area that was then, in the eyes of international law, vacant. The Cape had only effectively occupied Walfish Bay, which is, however, practically the only inlet and outlet for the trade of the country.

Reverting now to Natal, we find that it remained a province of the Cape until 1856, but in that year was formed into a distinct colony, under a Lieut.-Governor and a Legislative Council, a body in which the proportion of elective and non-elective members has been changed no less than six times, until, in 1893, responsible government was conceded to the colony.

From the outset, the claims of the natives—in spite of the "earth hunger" of the Boer settlers—to lands which they either held or occupied were scrupulously respected; with the result that, since the British occupation of the country, there has only been one serious trouble with the natives—the rebellion of Langalibalele,* which terminated in the banishment of the chief and the breaking up of the Hlubi tribe. With this exception, the natives of Natal have been under British rule, as Lord Wolseley reported, "happy and prosperous, well-off in every sense, and on the best of terms with the colonists." But for the sugar, arrowroot, and other growing industries on the coastlands, native labour could not be depended upon; "Coolies," or labourers from India, were therefore

* "The Great Sun which shines and burns."

introduced in 1860; and since then many thousands of Her Majesty's Indian subjects have settled down in the Colony, instead of returning home on the expiration of their contracts. Owing to the teeming native population and the introduction of coolies, there has not been any considerable influx of Europeans since Byrne's crude and ill-managed emigration scheme attracted some 4,000 British emigrants into the colony in 1848-51. There are now about 50,000 whites, the same number of coolies, and half a million Kaffirs.

When Natal was declared a distinct colony in 1856, serious troubles broke out in Zululand between King Panda's eldest son, Cetywayo, and his younger and apparently more favoured brother Umbalazi, which culminated in a terrible battle on the banks of the Tugela, in which Umbalazi and thousands of his followers were killed. Cetywayo thereupon became the real ruler of the country, and in 1861 was publicly announced as the future king. On the death of Panda in 1872, Cetywayo was installed as king of the Zulu nation by Mr. Shepstone, who, with his escort, was everywhere welcomed by the natives.

In the meantime, there had been almost continual disturbances in the Transvaal. President Burger's grandiose schemes for vivifying the Republic had come to nothing beyond driving the Boers into latent revolt—some indeed trekked away across the terrible Thirstland into Benguela, their path marked by a line of graves—and rendering the natives defiant and indeed uncontrollable. A campaign against the rebellious Bapedi, under their chief Sekukuni, turned out disastrous to the Dutch, who were also threatened by Cetywayo, who evidently wished to pose as a second Tshaka. Alarmed lest the excitement among the natives should spread and involve the colonies in danger, the British Government commissioned Sir Theophilus Shepstone to proceed to the Transvaal, and, if possible, to help the Boers out of their difficulties. He found the country in a state of anarchy, faith in the President gone, and his government defied; the people no longer willing to fight or to pay any taxes; the natives triumphant, and the country liable at any moment to be overrun by the impis of Cetywayo. An "emergency" had arisen, so pressing indeed that, to save the country, Shepstone, on the 12th of April, 1877, proclaimed it British territory, thereby transferring to the British the perplexing difficulties with the natives that had threatened to overwhelm the Dutch, and that taxed severely even our resources to overcome.

At this time an able Indian administrator—Sir Bartle Frere—became Governor of the Cape and High Commissioner for South Africa. After suppressing a rebellion of the Gcalekas and Gaikas in Kaffirland, he turned his attention to the critical position of affairs in Natal and the Transvaal. He found that Cetywayo had been allowed to develop the military system of the Zulus to an alarming extent. Sekukuni defied the British as he had the Dutch; but Sir Garnet Wolseley subdued the tribe, and took the bold chieftain prisoner. The award in the Zulu-Dutch frontier dispute with Cetywayo was in favour of the Zulus, but the Zulu king's disposition was so hostile that, in December, 1878, along with the boundary award, an ultimatum was sent him—requiring

him to disband his regiments and to give satisfactory assurances for the peace and quiet of his country. Cetywayo refused, and on the 10th of January, 1876, the English army advanced unopposed into Zululand in three divisions. Ten days later the centre column, under Lord Chelmsford, encamped at the foot of Isandlwana—the hill of "the little hand;" but though there were waggons enough to form a laager, none was made, nor was a trench dug. At dawn, on the 22nd, part of the column marched to attack a kraal some miles distant, and while these troops were away, ten or eleven of Cetywayo's regiments, in all about 23,000 or 24,000 men, suddenly surrounded the camp, and massacred nearly 700 British soldiers and 130 colonists. Very few escaped. Lieutenants Melville and Coghill gallantly endeavoured to save the colours of the 24th regiment, but were both shot. The Zulus suffered severely; three thousand of them were killed in the desperate fight for life on the "Flodden" of Natal. About five o'clock on the same day, some 4,000 Zulus attacked the depôt and hospital post at Rorke's Drift, and, until four o'clock the next morning, the little garrison, behind a slender barrier of sacks of maize and of biscuit boxes, repulsed the fierce assaults of the Zulus. This splendid defence no doubt saved Natal from a serious invasion. Lord Chelmsford encamped that night on the fatal field of Isandlwana, and then retreated into Natal. Strong reinforcements soon arrived, and another advance was made; and at Ulundi, on the 4th of July, Cetywayo's impis made their last stand. With magnificent courage the Zulu warriors rushed on the British square, but were literally mown down by the terrible hail of bullets; and, turning to retire, were charged by the British cavalry and dispersed, never again to rally. Cetywayo fled, but was soon captured, and sent a prisoner to Cape Town. During the war, the hapless Prince Imperial of France, while out with a small reconnoitring party, was surprised and killed by a band of Zulus. Sir Garnet Wolseley parcelled out the country between thirteen kinglets, all of whom he placed under the control of a British Resident. This arrangement did not work well, and in 1883, Cetywayo, who in the meantime had visited England, was restored to part of his former dominions. Another portion was formed into a "Reserve" for those who did not wish to be under the king, while a small territory was left to Sibepu, who shortly after attacked Cetywayo and forced him to take refuge in the Reserve, where he died, or some say was poisoned, the year following. The implacable Sibepu continued to fight Cetywayo's son and successor, Dinizulu, who called to his aid a number of Boer farmers, whom he rewarded with a large tract of land in Western Zululand, which was then formed into the "New Republic," and is now a part of the Transvaal. Sibepu was subdued, but disturbances continued, and in 1887 Zululand was formally annexed and declared a British Crown Colony. Dinizulu was naturally indignant, and headed a revolt against British authority, but was arrested, and, along with the other chiefs, exiled to St. Helena. Sir Marshall Clarke was appointed Resident Commissioner in 1893, and his efforts will no doubt be crowned with the same success in Zululand as they have been in Basutoland.

The close interdependence of events in Natal and Zululand has its counterpart in the still closer connection between the history of the Orange Free State and that of Basutoland. When British sovereignty over the Orange River territory was withdrawn, a small and scattered community of farmers was left to set up and maintain a government of its own, while, close by, a powerful and hostile native state had been created by the genius of the astute and sagacious Basuto chieftain, Moshesh, undoubtedly the ablest black ruler that South Africa has ever produced. Secure in his impregnable mountain-fortress of Thaba Bossigo, Moshesh rallied round him the wretched remnants of the Bechuana tribes decimated by the savage impis of Tshaka, and by his clemency attached to him even those whom war and famine had caused to become cannibals. He knew how to change foes into friends; and he put an end to the raids of the Matabeles by a most un-African proceeding. He welcomed the missionaries, and "admired the white people," so long as they did not thwart his plans. He allowed some of the emigrant farmers to settle on Basuto territory —"they might remain for years if they liked." Under his wise and kindly rule, the Basutos increased so rapidly that they wanted more land, and Moshesh re-claimed the farms occupied by the Boers. "He had lent them the cow to milk; they could use her, but they could not sell the cow." Thus originated an endless series of boundary disputes, which involved the Basutos in a long and bitter war with the Free State. The light horsemen of Moshesh ravaged the Free State farms and then retreated into their fortified caves and mountain strongholds. The farmers retaliated as best they could. For ten years the fighting went on, but in March, 1868, just as the last Basuto stronghold was on the point of surrendering, Moshesh transferred the sovereignty of his country to the Queen. To the surprise and disgust of the Free State burghers, Sir Philip Wodehouse declared the Basutos British subjects, and sent an armed force to protect them. Peace was concluded in 1869, and, in 1871, Basutoland was annexed to the Cape. Nine years later, the Cape Government attempted to disarm the Basutos, but they resisted so strenuously that the colonial forces absolutely failed to reduce them to submission. A disannexation Bill was therefore passed in 1883, and the year following, in compliance with the request of the Basutos themselves, the country was re-transferred to the Imperial Government, and has since been ruled by hereditary chiefs under the direction of a British administrator and magistrates.

In the meantime the Free State had been progressing, slowly but surely, when in a corner of what Sir George Clerk had called a "howling wilderness," between the Modder and the Vaal, a discovery was made that created almost a revolution in South African affairs. The Diamond Fields attracted thousands of adventurers from all parts, and there was naturally much confusion and lawlessness. The Free State sent its officers to govern the district, but the Griqua captain Waterboer and the South African Republic claimed the ground. The matter was submitted to arbitration, and as soon as the "Keate award," which was in favour of Waterboer's claims, was issued, Sir Henry Barclay proclaimed the Griqua captain's

country a British dependency, and formed Griqualand West into a British Crown Colony. The Free State, however, protested; and, as it was afterwards found that Waterboer had really no right to the territory, the British Government paid the republic £90,000 as a solatium, and offered £15,000 more to encourage the construction of railways in the country. For 20 years, however, the ox-waggon remained the only means of transport; but, in 1892, the main trunk line from Cape Town to Pretoria, which passes through the State, was opened, while branch lines connecting with Natal and the Eastern system of the Cape were also constructed. Griqualand West and the Diamond Fields, which thus justly belong to the Free State, became, in 1880, an integral part of the Cape Colony.

When the Transvaal was annexed in 1877, the President retired under protest, and there was considerable dissatisfaction among the Boers; but as long as Sir Theophilus Shepstone, who was personally much liked by them, remained at the head of affairs, there was no open opposition to English rule. One deputation after another visited England to protest against the annexation, and to endeavour to get it annulled; but the Boer delegates only received a decided "No" in answer to their earnest appeals, and Sir Garnet Wolseley told the Boers at Pretoria that, "so long as the sun shone in the heavens," so long would the Transvaal remain English territory. The appointment of Sir Owen Lanyon to succeed a popular man like Shepstone, and the promulgation of a so-called "constitution" for the country—a "nominated" mockery of the freely-elected Volksraad—hastened the crisis; and at a great meeting at Paarde Kraal, where the gold-mining town of Krugersdorp now stands, the Boers resolved to fight for their independence, and, if beaten, to burn their homesteads, lay waste the country, and trek north beyond the Limpopo. Three commandos were immediately formed; and at Heidelberg, on the historic 16th of December—Dingaan's Day—the flag of the republic was again hoisted amid enthusiastic cheering.

The brief war which followed was disastrous to the British arms. Pretoria and other towns, garrisoned by English troops and crowded with English settlers and loyal Transvaalers, were surrounded and their communications cut off, while a detachment of the 94th was almost annihilated at Bronkhorst Spruit by a party of mounted Boers. The excitement in the neighbouring English colony of Natal was intense, and General Sir George Colley, the Governor, hastily formed a relief column of about 1,000 men, and at once marched north. Immediately a much stronger force of Boers crossed the border into Natal, and took up a strong position on both sides of the road which winds through the narrow pass of Lang's Nek, and along which the relief column would have to pass. A determined attempt to force a passage was repulsed with heavy loss by the deadly fire of the concealed Boer marksmen. Some days later, an English patrol of about 300 men was attacked on the Ingogo Heights and forced to retreat, leaving two-thirds of their number dead or wounded. On Saturday night, the 26th of February, Sir George Colley left the camp at Mount Prospect with 600 men, and silently climbed to the top of Majuba Hill. At dawn the next day, the

Dutchmen encamped at the Nek, two thousand feet below, were astonished to see the redcoats on the heights above them, and prepared to retire. Commandant-General Joubert called for volunteers to storm the apparently impregnable British position. A hundred and fifty young Boers responded, and, firing continuously, crept up from terrace to terrace. At noon, about seventy of them reached the summit and fired a deadly volley into the terrified soldiers, who broke and fled down the hill, leaving 94 killed, 134 wounded, and 57 prisoners. This daring deed was done with a loss to the Boers of only one man killed and five wounded. General Colley himself was among the slain, and but for the Highlanders, who were entrenched on a connecting ridge, very few indeed of the British soldiers would have escaped. Commandant Joubert reported to Kruger that "the troops fought like true heroes, but God gave us the victory."

This fatal fight ended the war. Reinforcements poured into Natal; General Roberts was sent out to command the avengers, but he was stopped at Cape Town, and Sir Evelyn Wood received orders to conclude an armistice and arrange terms of peace. Complete self-government was restored to the Boers, subject to the suzerainty of the Queen. The Transvaal was no longer English territory, and yet "the sun still shone in the heavens."

A convention, embodying the terms of peace, was ratified by the Volksraad in 1881, and was subsequently modified, the only vestige of British control being the power of vetoing any treaties the Republic may make with any State or nation other than the Free State. It also provided for the pacification of the western borderlands, where Massouw and Moshette, two chiefs allied with the Boers, were pitted against Mankoroane and Montsioa, two rival chiefs who had, of course, sided with the English during the war. European freebooters were enlisted by both parties, and rewarded with tracts of land, which were forthwith formed into two Republics—Stellaland and Goshen—which incontinently dissolved on the approach of Sir Charles Warren's expedition. There was no fighting, but it showed the natives and the Boers that the British Government was not to be trifled with, and it kept open the great highway into the interior. In 1885, British authority was formally proclaimed over the whole of Bechuanaland, the southern section being formed into a Crown Colony and placed under British magistrates, while in the northern division the chiefs were permitted to continue to exercise full authority over their people. Of these chiefs, Khama, the wise and able head of the Bamangwatos, is the best known and the most respected. His kindliness and generous treatment of travellers and traders are proverbial, but he will not permit intoxicants to be brought into his country, and his town, Palapye, is the model native town of South Africa.

The Republic had not long been re-established when the painful financial troubles of the Burger's régime recommenced, and the country was rapidly drifting from bad to worse. Dissension among the leaders, discontent among the burghers, the Treasury in sore straits for money which could not be got, the future of the country seemed dark indeed, when the discovery of the richest goldfields in the world saved the State from bankruptcy, and filled

to overflowing its empty coffers. Dazzling descriptions of the Sheba Reef, in the De Kaap valley, and almost incredible accounts of the marvellous richness of the Witwatersrand "banket," attracted thousands of miners and capitalists, artisans and traders, into the country. A busy town sprang up in the centre of the eastern mines, but was soon eclipsed by another, which grew with still more wonderful rapidity into a populous city, with long streets of handsome buildings, numerous suburbs, and quite a number of outlying towns. English enterprise and capital soon wrought a marvellous change ; trade revived, and the Government got a new lease of vigorous life.

The relations between the Transvaal authorities and the British Government—which, in 1885, had extended its sovereignty over British Bechuanaland and the country northward to the Zambesi, and had, in 1888, concluded a treaty of peace and amity with Lo Benguln, the king of Matabeleland and overlord of Mashonaland—were considerably disturbed, in 1891, by rumours of a great Boer trek into Mashonaland, which had been occupied by the British South Africa Company, under a royal charter granted in 1889. The Company's "Pioneer Expedition" crossed the Macloutsie River in June, 1890, and reached Mount Hampden, in Mashonaland, on September 12th, without the loss of a single man. A fort was built at Tuli drift ; another at Victoria, on the edge of the high Mashona plateau, and not far from the wonderful ruins of Zimbabwe ; and two more at Charter and Salisbury, the latter, near Mount Hampden, being made the headquarters of the administration for Mashonaland.

Meanwhile the proposed Boer trek had been effectually damped by "Oom Paul," who, however, expected to be allowed in return to annex the Swazi country to the Republic—a step which the Swazi queen and people do not devoutly wish. About a hundred armed Boers appeared on the banks of the Limpopo, but found a strong body of the Company's police stationed at the drifts. The Boer leaders crossed and were arrested ; a few of the trekkers accepted the Company's terms, and were allowed in return to enter the country ; the rest "sold or bartered what they had to the Company's commissariat, and returned home sadder if not wiser men"

The Company did not limit their sphere of operations to Mashonaland and the country to the south of the Zambesi. Mr. Rhodes sent missions to Lewanika, chief of the Barotse, and to other chiefs between the Barotse country and the Nyasaland Protectorate. In all these territories, as well as in Gazaland, in Portuguese East Africa, valuable trading and mining concessions have been secured for future exploitation. The total extent of territory, to the north and south of the Zambesi, within the sphere of the British South Africa Company, is estimated at over three-quarters of a million square miles, an area more than six times that of Great Britain and Ireland.

The Protectorate of Nyasaland has been placed under an Imperial Commissioner, who also acts as Administrator of the British South Africa Company's trans-Zambesian sphere of operations. He resides at Zomba, on the healthy and fertile uplands of the Shiré ; and with a land force of Sikhs, supported by gunboats on the Shiré and Lake Nyasa, has done much towards suppressing the slave trade, and ensuring the peaceful development of the country.

The Company's right to enter Mashonaland was based on the Rudd-Rhodes concession, granted by Lo Benguln, who, however, would not allow any considerable number of white people to enter into his own country, and was naturally jealous of the Company's occupation of Mashonaland. The greatest possible care was taken to prevent collisions with the Matabele, and the route of the Pioneer Expedition was purposely planned to avoid their kraals. But the Matabele warriors persisted in making raids on the wretched Mashonas, and even attacked the Mashona employés of the Company, until at last hostilities broke out. A well-equipped force of about 600 men advanced westwards from Forts Charter and Victoria, while another strong force of the Company's and Bechuanaland Border Police, with about 1,500 Bamangwatos under Khama, advanced from the south. Khama soon withdrew his men, but the European forces marched towards Buluwayo, and, after repulsing several desperate attacks, entered the Matabele capital. Lo Benguln's power was effectually broken, and he fled with a number of warriors towards the Zambesi, closely pursued by a small patrol, which was, however, cut to pieces on the banks of the flooded Shangani River. Before leaving, Lo Benguln had given orders to burn Buluwayo, and the victorious troops found the place in flames, but a new township soon sprang up around it, and it now forms the seat of government for the country, and the chief centre of its trade and industry. The great attraction of Mashonaland and Matabeleland consists in the gold-bearing reefs, of more or less richness, which traverse both countries in all directions, and many of which are being actively developed. Rapid progress is, however, impossible, until the means of communication have been improved and extended. The "fly country" between the Mashona uplands and the coast has already been spanned by the Beira Railway, and an excellent waggon road leads from the terminus, via Umtali, to Salisbury and other centres. The Cape-Kimberley Railway is being extended northward from Vryburg, and this line, or the Cape-Pretoria line, will no doubt ultimately be extended north to the Tati and Buluwayo. The rapid extension of the railways within the last five years has marked the commencement of a new era in South Africa—an era, let us hope, of concord between Bantu, Boer, and Briton ; an era of peace, prosperity, and progress.

GEORGE PHILIP AND SON, LONDON AND LIVERPOOL.

TABLE OF DISTANCES IN NAUTICAL MILES.

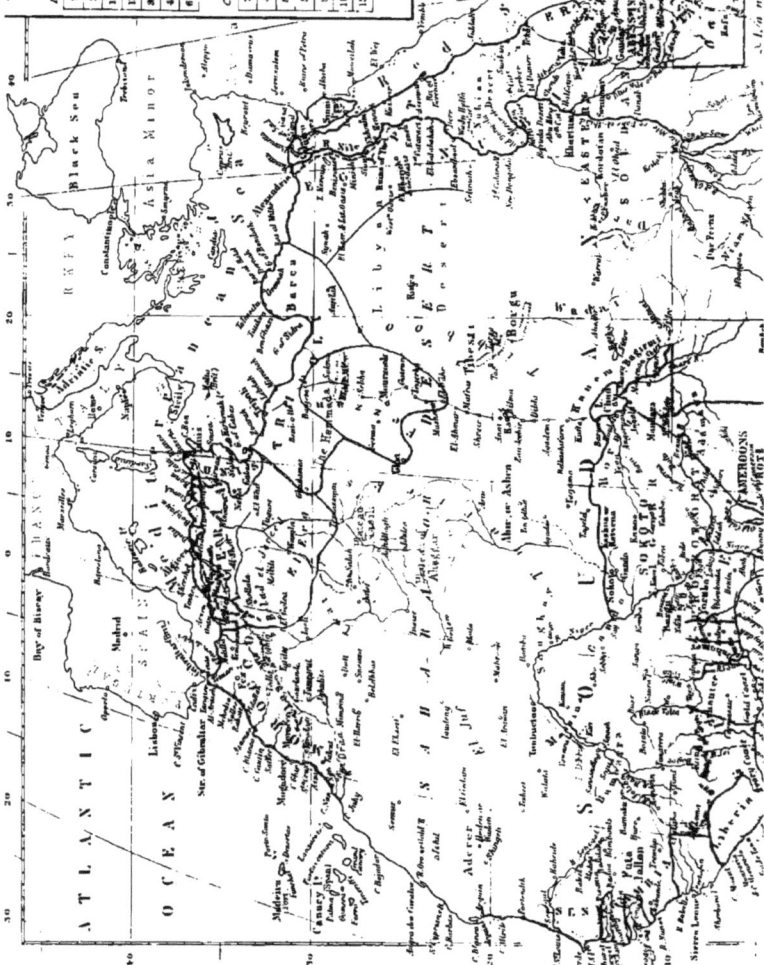

Atlantic Distances.

London.	Southampton.	Plymouth.	Madeira.	Grand Canary.	Ascension.	St. Helena.	Cape Town.
202	197						
317	127	1398	1230				
1420	1304	1180	7400	269			
1714	1810	1813	3820	3461	2209		
3009	8713	4203	5061	807			
4809	4311	8092	4413	3266	1708		
6181	5078						

Coast Distances.

Cape Town.	Mossel Bay.	Algoa Bay.	Port Alfred.	East London.	Natal.	Delagoa Bay.	Beira.
262	145						
805	259	70					
579	317	131	90				
912	670	284	239	203			
1112	870	644	611	533	300		
1367	1325	1100	1006	1024	797	495	

Natal to Tamatave, 1363.
Tamatave to Mauritius, 460.

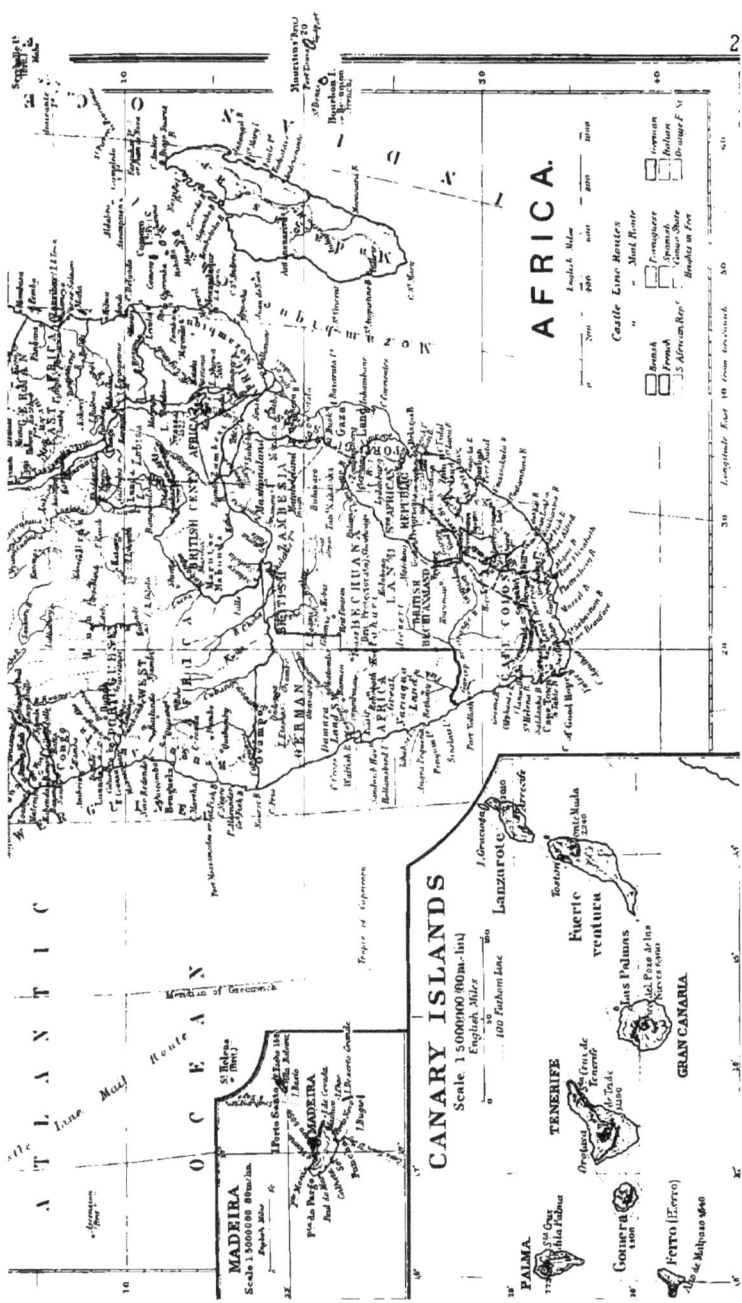

AFRICA.

Castle Line Routes
— Mail Route

English Miles
0   200   400   600   800   1000

Longitude East 10 from Greenwich

ATLANTIC OCEANN

Meridian of Greenwich

Castle Line Mail Route

CANARY ISLANDS

Scale 1:5,000,000 (80 m. 1 in.)

English Miles
0   10   30   50

100 Fathom Line

Lanzarote

J. Graciosa

Fuerteventura

Tostou

Gran Canaria

Las Palmas

TENERIFE

Santa Cruz de Tenerife

Orotava

Pico de Teide

PALMA

Santa Cruz de la Palma

Gomera

Ferro (Hierro)

MADEIRA

Scale 1:5,000,000 (80 m. 1 in.)

English Miles

Porto Santo

St Helena

Tropic of Capricorn

GERMAN EAST AFRICA

BRITISH CENTRAL AFRICA

ZAMBESIA

GERMAN SOUTH WEST AFRICA

BRITISH BECHUANALAND

CAPE COLONY

C. of Good Hope

Walfish Bay

Port Nolloth

Cape Town

Port Elizabeth

C. Verde

CONGO

Cameroon

Gold Coast

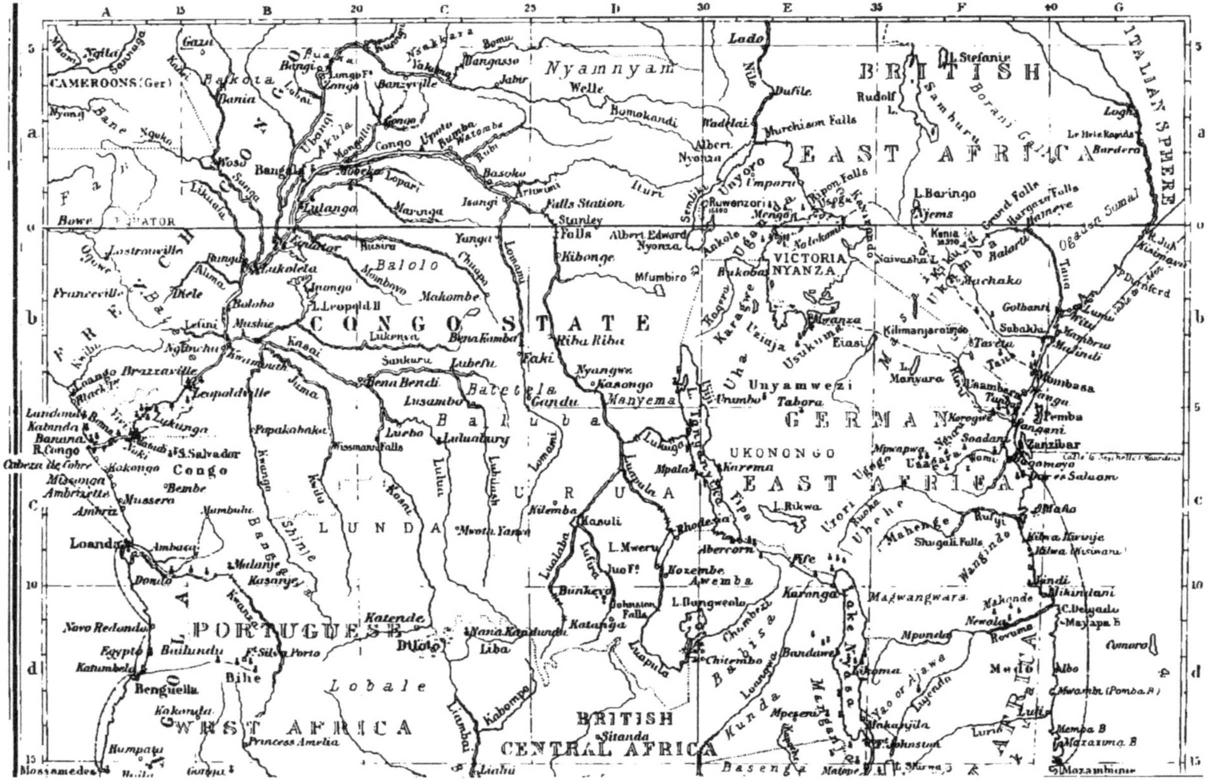

ITALIAN SPHERE
BRITISH EAST AFRICA
CONGO STATE
GERMAN EAST AFRICA
BRITISH CENTRAL AFRICA
PORTUGUESE WEST AFRICA
FRENCH CONGO
CAMEROONS (Ger)
VICTORIA NYANZA
Nyamnyam
UKONONGO
LUNDA
URUA
Baluba
Balolo
Lobale
Lake Nyasa
Tanganyika
L. Rudolf
L. Stefanie
L. Baringo
Albert Nyanza
Albert Edward Nyanza
L. Mweru
L. Bangweolo
L. Rikwa
Zanzibar
Mombasa
Loanda
Benguella
Lado
Nile
EQUATOR

CENTRAL & SOUTH
AFRICA.

Showing the
COMMUNICATIONS & MISSION STATIONS.

British Territory Coloured Red.

SCALE 1:17,500,000
English Miles

Railways open
constructing or projected
Coach Routes
Telegraphs & Cables

• Mission Stations
"Castle Line" Route
Navigable Rivers
Rapids & Falls

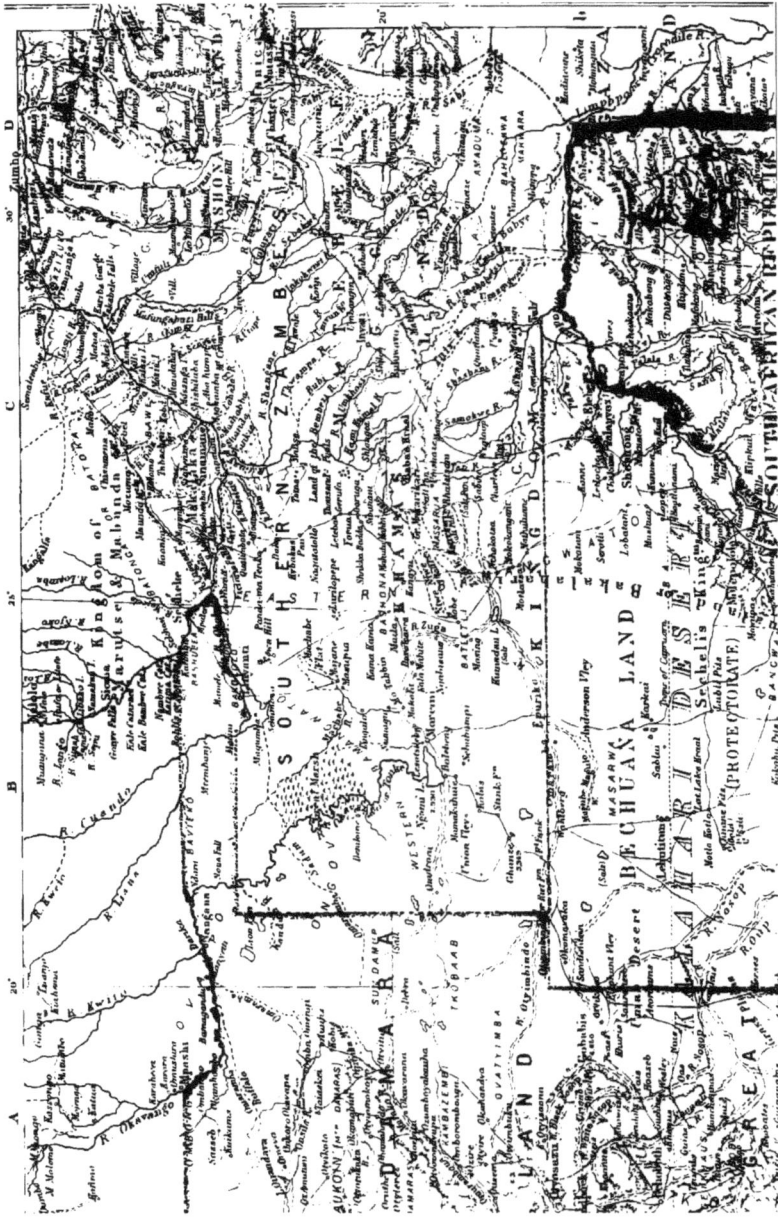

AFRICA

SOUTH OF THE ZAMBESI.

POLITICAL & INDUSTRIAL

SCALE 1:7,500,000 (120 Miles = 1 Inch).

British Territory — South African Republic
Portuguese " — Orange Free State
German " — Copper — Coal

◎ Diamonds ○ Gold ○ Copper □ Coal
O=Ostrich S=Sugar W=Wool C=Cattle, F=Fruit, G=Grain, Mz=Maize, Wn=Wine, Wh=Wheat

☐ Steppe and Desert Regions. ☐ Grass and Arable Lands.

TRANSVAAL

BECHUANALAND (CROWN COLONY)

NAMAQUA LAND

CAPE COLONY

CAPE TOWN
Cape of Good Hope

Long. East 25° of Greenwich

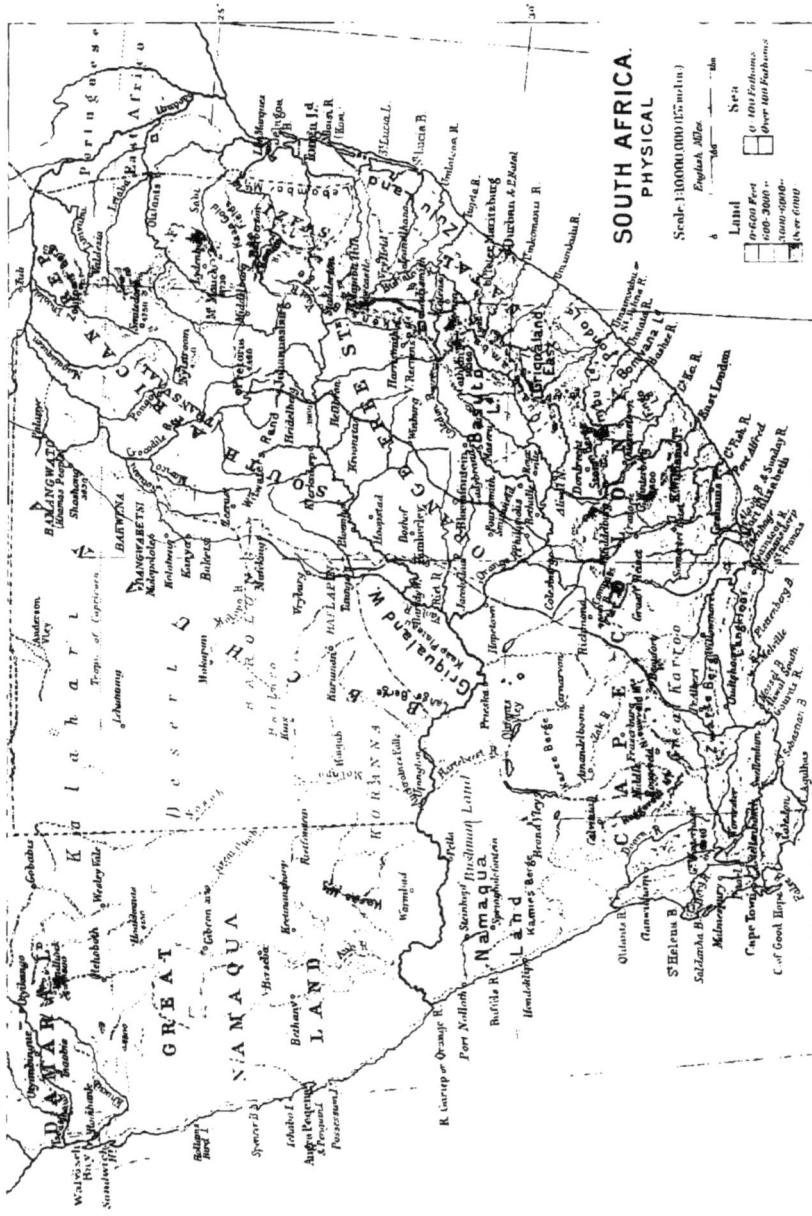

SOUTH AFRICA.
PHYSICAL

Scale 1:10,000,000 (157 m.in.)

English Miles

Land		Sea	
0-600 Feet		0-100 Fathoms	
600-3000 "		over 100 Fathoms	
3000-6000 "			
over 6000 "			

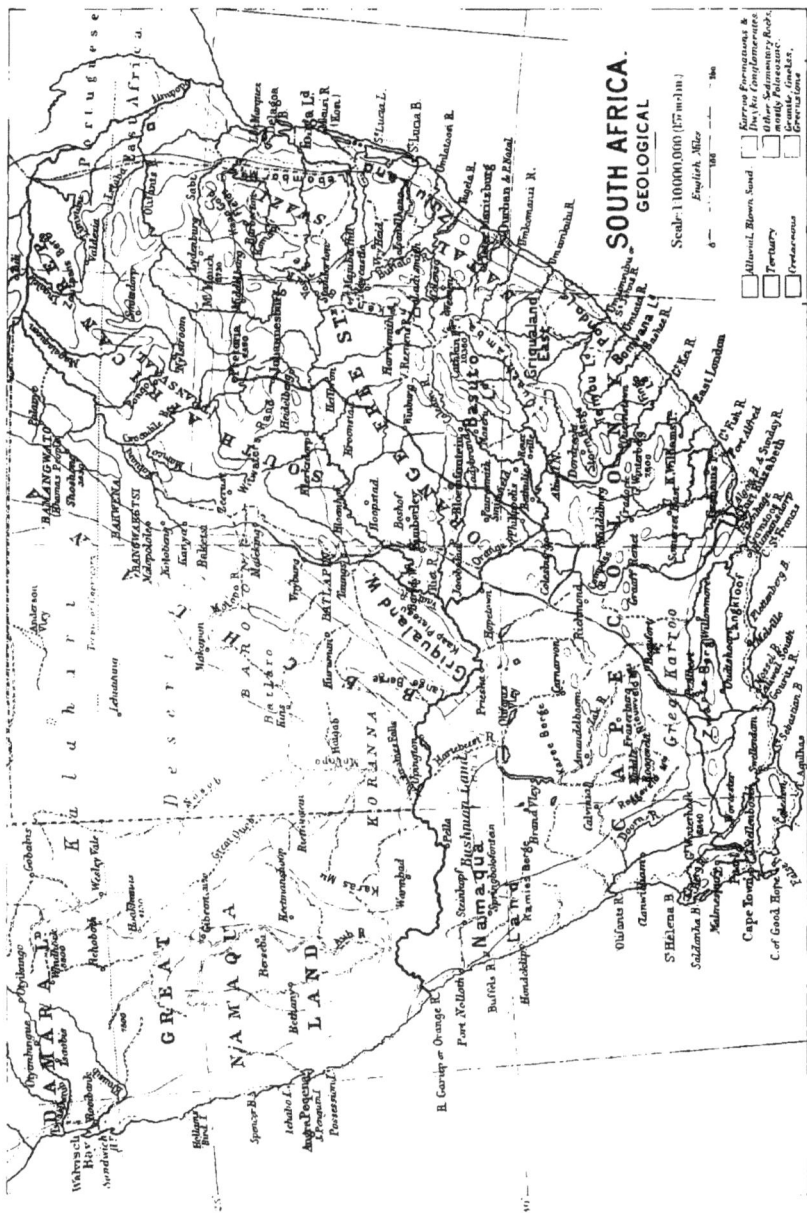

SOUTH AFRICA.
GEOLOGICAL.

Scale 1:10,000,000 (157 m. lin.)

English Mile

	Alluvial, Blown, Sand	Karroo Formations & Beds, Conglomerates
	Tertiary	Other Sedimentary Rocks, mostly Palæozoic
	Cretaceous	Granite, Gneiss, Greenstone

.

SOUTH AFRICA
RAINFALL

Scale: 1:10,000,000 (157 m.=1 in.)

English Miles

Under 10 inches
10 - 25 inches
Over 25 inches

The figures denote the
annual Rainfall in inches

6

SOUTH AFRICA
ETHNOGRAPHICAL

Scale 1:10000,000 (157m=1in.)

English Miles

The Whites number over 50 per cent
of total Population
25-50 per cent
less than 25 per cent
West of this line Hottentots & Bushmen
form the majority of the population

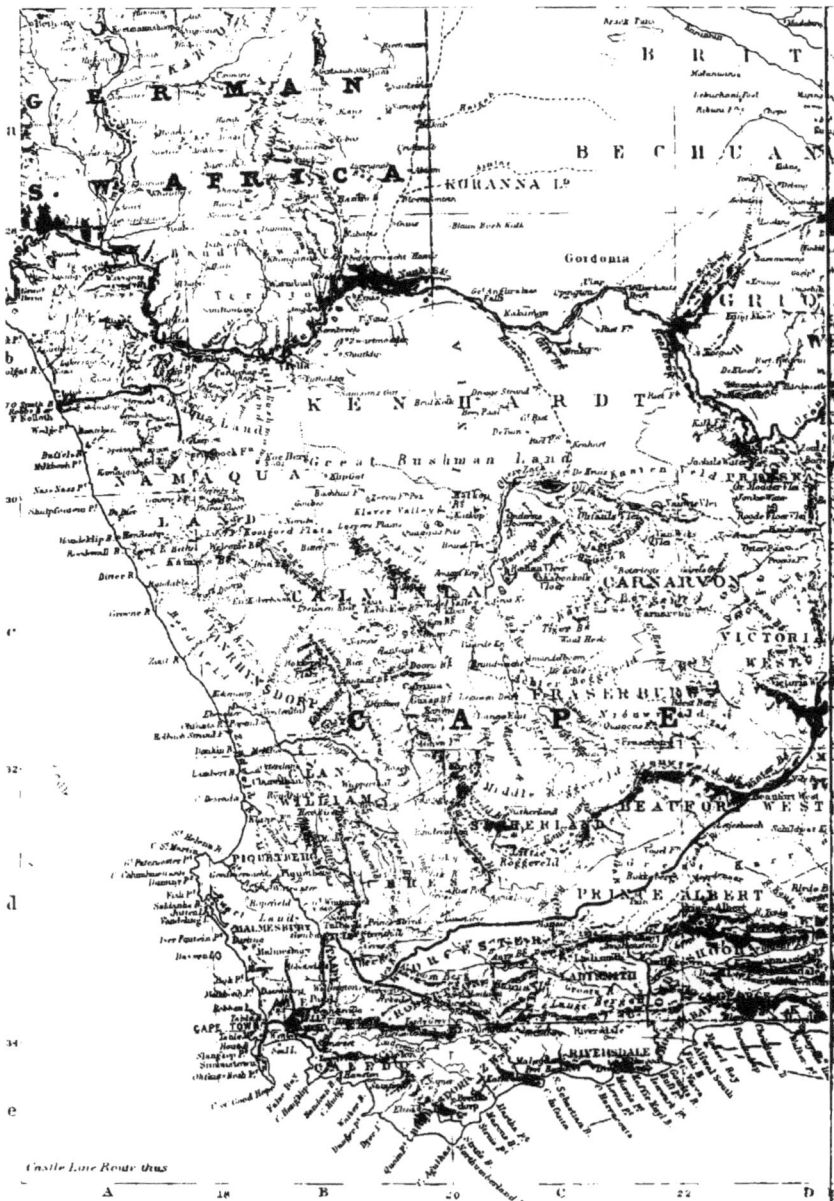

CAPE COLONY

IN DIVISIONS

with

BASUTOLAND & THE ORANGE FREE STATE

SCALE 1 : 4 400 000 (72 M = 1 Inch.

Statute Miles

0   10   20   40   60   80

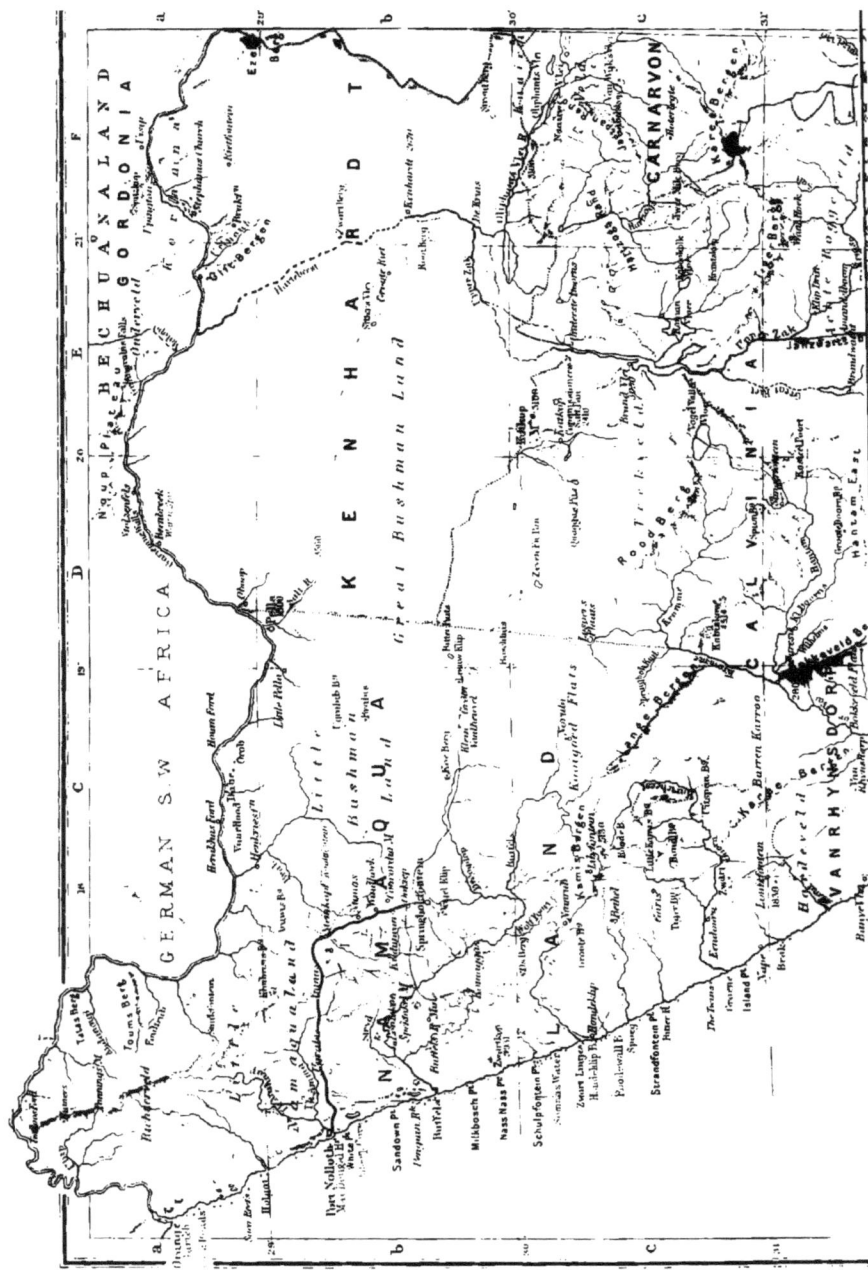

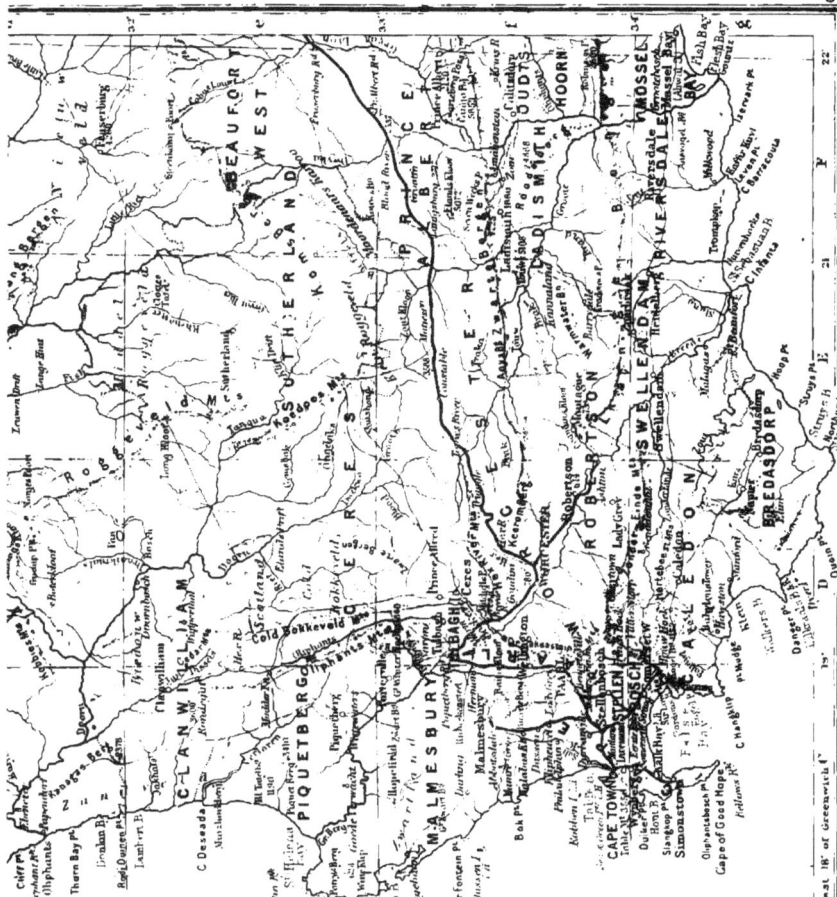

SOUTH AFRICA
WESTERN SHEET

SCALE 1:2,500,000 (40 MILES=1 INCH)

English Miles

Railways open —— Roads —— Rivers
constructing —— 8 Mission Stations
Heights in feet

Long. East 18° of Greenwich 17°

T   L   A   N   T   I   C

O   C   E   A   N

BEAUFORT WEST

PRINCE ALBERT

CLANWILLIAM

CERES

MALMESBURY

PIQUETBERG

TULBAGH

WORCESTER

ROBERTSON

SWELLENDAM

CALEDON

BREDASDORP

MOSSEL BAY

OUDTSHOORN

CAPE TOWN

STELLENBOSCH

SIMONSTOWN

Cape of Good Hope

SOUTH AFRICA
CENTRAL SHEET
SCALE 1:2500000/40 MILES=1INCH
English Miles

Railways open — Roads n Passes
constructing & Mission Stations
Heights in feet

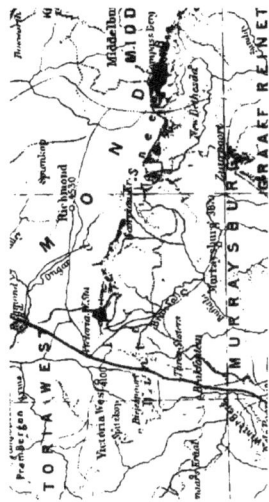

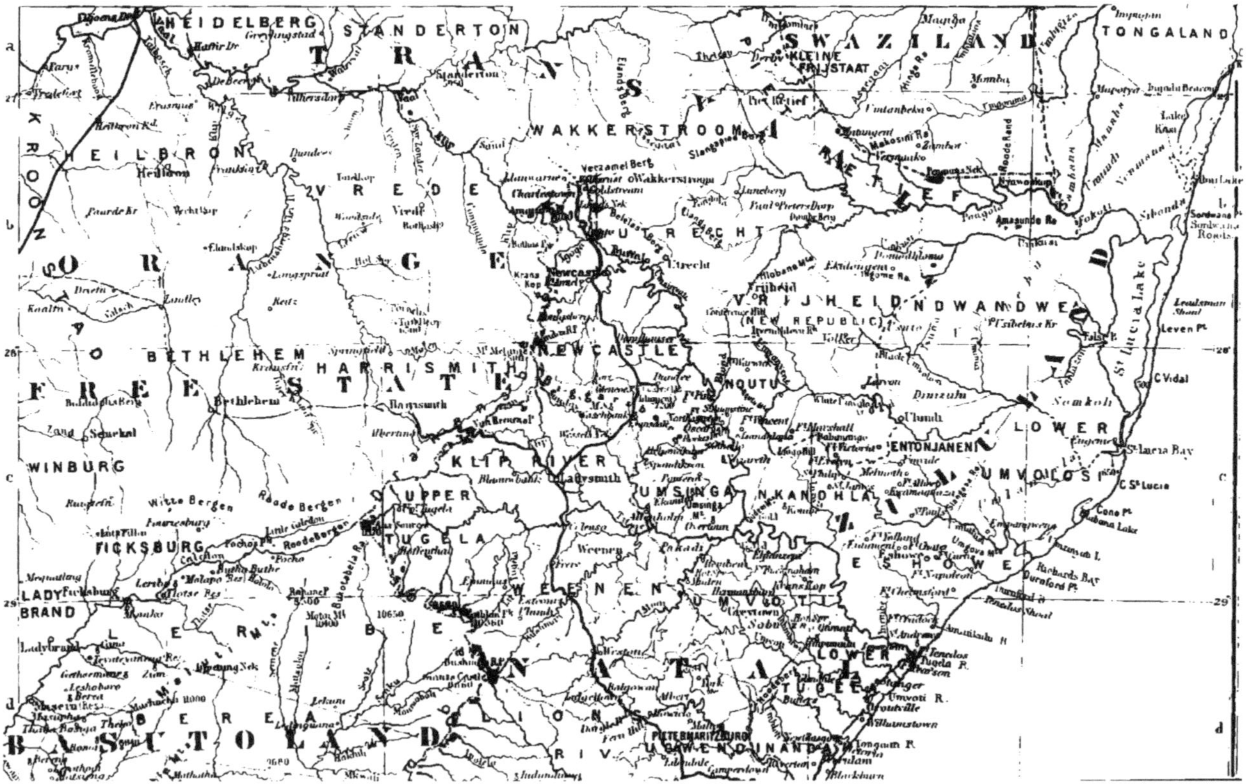
HEIDELBERG
STANDERTON
TRANSVAAL
SWAZILAND
KLEINE VRIJSTAAT
TONGALAND
WAKKERSTROOM
HEILBRON
VREDE
UTRECHT
ORANGE FREE STATE
VRIJHEID
NDWANDWE
(NEW REPUBLIC)
BETHLEHEM
HARRISMITH
NEWCASTLE
NQUTU
ZULULAND
LOWER UMVOLOSI
ENTONJANENI
WINBURG
KLIP RIVER
UMSINGA
NKANDHLA
UPPER TUGELA
FICKSBURG
ESHOWE
LADYBRAND
WEENEN
UMVOTI
LOWER TUGELA
NATAL
BEREA
BASUTOLAND
PIETERMARITZBURG
INANDA
St Lucia Lake
St Lucia Bay
C St Lucie
C Vidal
Leven Pt
Richards Bay
Ladysmith
Bethlehem
Harrismith

10

SOUTH AFRICA

EASTERN SHEET

including

NATAL, BASUTOLAND, ZULULAND, AND EASTERN CAPE COLONY.

SCALE 1:2,500,000 (40 MILES=1 INCH)

English Miles

—— Railways open   ···· Roads   △ Passes
···· Railways in progress   ⚑ Mission Stations
Heights in feet

PIETERMARITZBURG

Scale Feet

CAPE TOWN

1. Jewish Synagogue.
2. Dutch Reform Church.
3. St. Andrew's Church.
4. Congregational "
5. St. Mary.
6. St. Paul's School.
7. St. Paul's.
8. Public Works.
9. Drill Hall.
10. Old Somerset Hospital.
11. New Somerset Hospital.

12. Government Ho.
13. Houses of Parliament.
14. Museum.
15. St. George's Cath.
16. Roman Catholic Cath.
17. Baptist Church.
18. Dutch Reform Ch.
19. Town Hall.
20. Mint.
21. New Post Office.

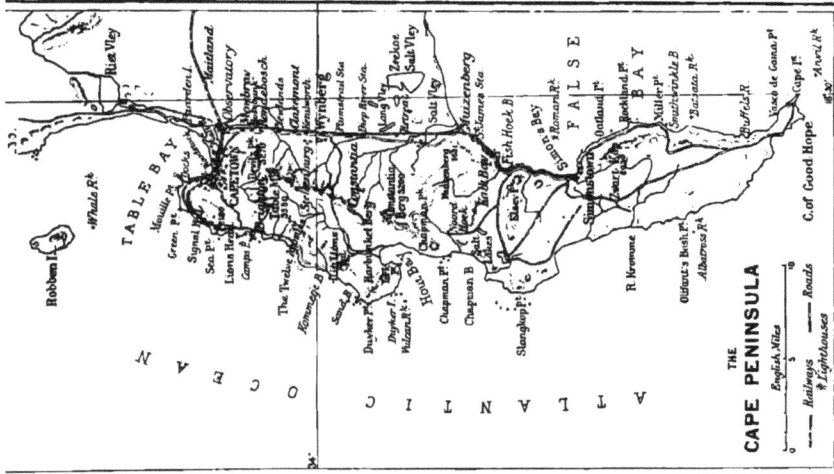

11

THE
CAPE PENINSULA

English Miles

Railways ——— Roads

☩ Lighthouses

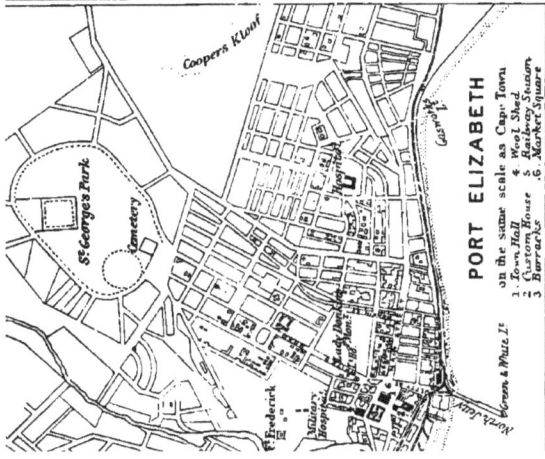

PORT ELIZABETH

on the same scale as Cape Town

1. Town Hall   4. Wool Shed
2. Custom House   5. Railway Station
3. Barracks   6. Market Square

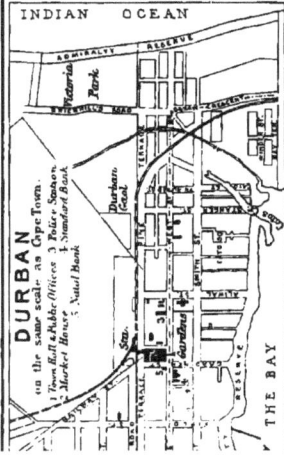

INDIAN OCEAN

DURBAN

on the same scale as Cape Town.

1. Town Hall & Public Offices   3. Police Station
2. Market House   4. Standard Bank
5. Natal Bank

THE BAY

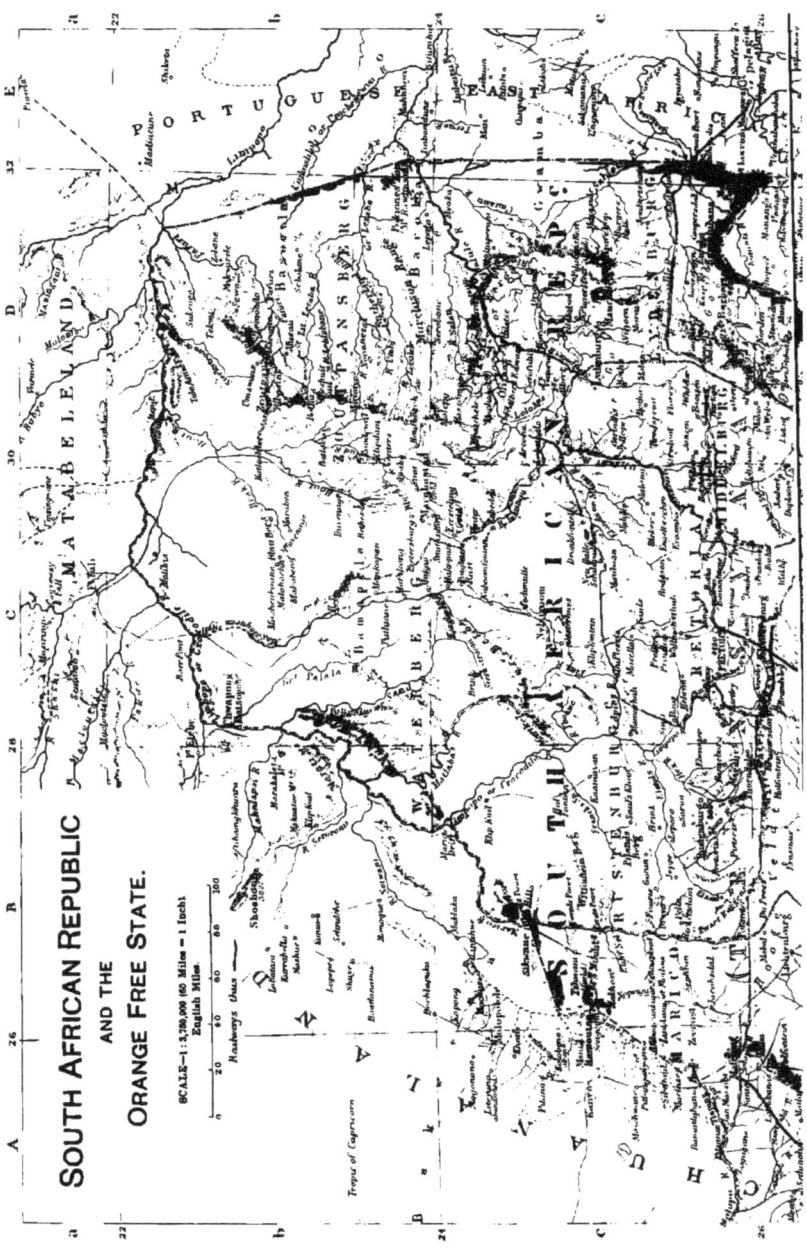

SOUTH AFRICAN REPUBLIC

AND THE

ORANGE FREE STATE.

SCALE=1:5,200,000 (60 Miles = 1 Inch)
English Miles

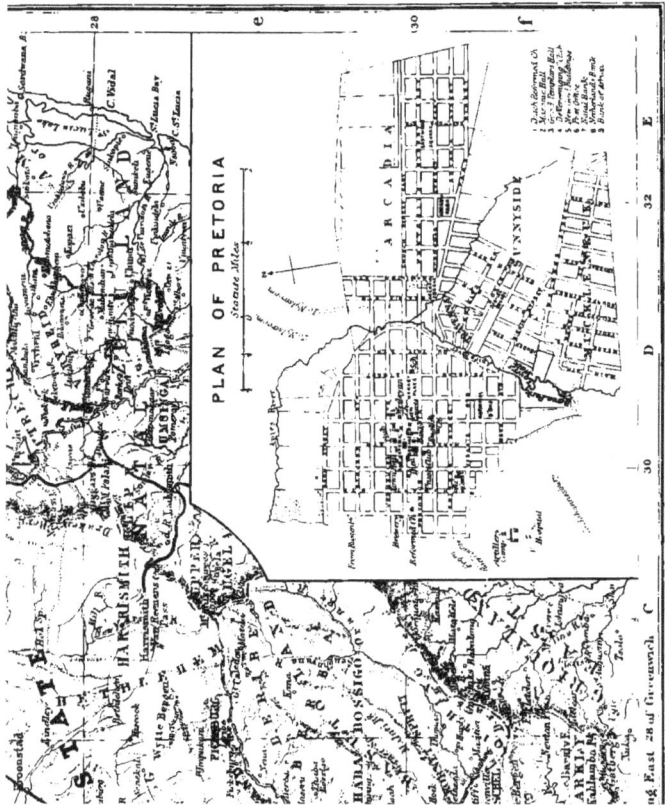

PLAN OF PRETORIA

JOHANNESBURG
AND ITS SUBURBS

Auckland Park
KLIPFONTEIN
Belleview
Yeoville
MIDDLEFONTEIN
Bertram
Doornfontein
Bramfontein
Troyeville
Paarls Hoop
Turffontein
Marshall
Ferreira Mine
DOORNFONTEIN
LANGLAAGTE
Natal Spruit
Prospect
Booysens
TURFFONTEIN
Ingram burg
Bezuidenville
CONCORDIA
Bellevue
Casey n
Rosettenville
VIERFONTEIN
13A
Stanute Miles

24'

W A T E[R]

Marikwe

Krans Berg

S O U

Witfontein Berge
Omars Berge
R U S T E N B U R G
Gr Bruk Spr
Mabies Kr
Kaulspoart
Eilands b
MARICO
R
Enzelsberg
Zeerust
Marico
Rustenburg
Zwartruggens
Bechuanaland
Mafeking
Lichtenburg
Doornkop
W
Erasmus
T
Lichtenburg
Tafelkop
Krugersdorp
Roodekop
Schulenburg
Ventersdorp
LICHTENBURG POTCHEFSTROOM
Goldfield
Potchefstroom
Klerksdorp
Ventersdorp
BLOEMHOF KROONSTAD
O R A N G E

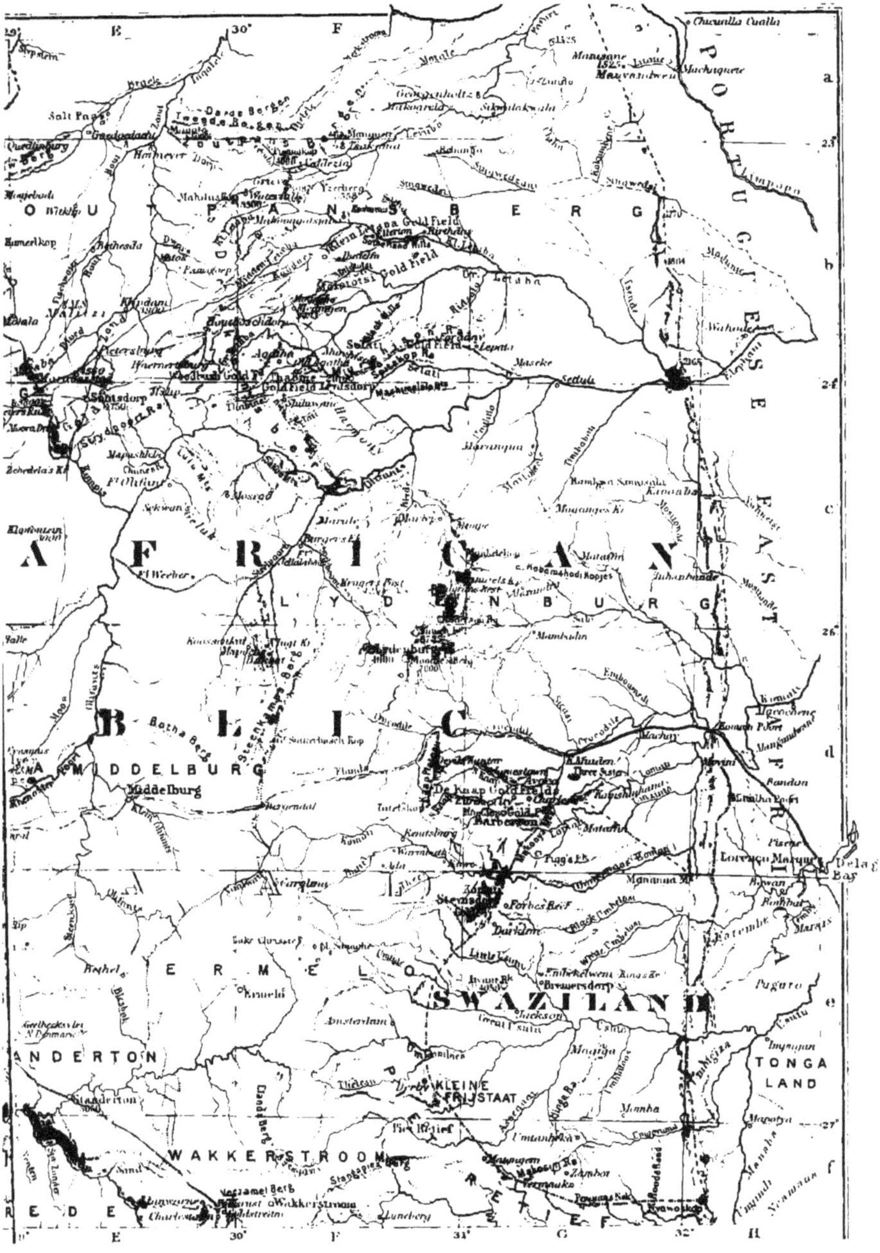

E
30°
F
G
31°
32°
H
a
b
c
d
e
f
Limpopo
Salt Pans
Quedlinburg
Sand
Letaba
Selati
Klein Letaba Gold Field
Selati Gold Field
Leydsdorp
Olifants
Smitsdorp
Pietersburg
Woodbush Gold Fd.
Komati Poort
Middelburg
MIDDELBURG
Botha Berg
Steenkamps Berg
Lydenburg
L Y D E N B U R G
Krugers Post
Barberton
De Kaap Gold Fields
Forbes Reef
Darkton
Carolina
Bethel
E R M E L O
Ermelo
Amsterdam
Jackson
Bremersdorp
SWAZILAND
KLEINE
FRIJSTAAT
Piet Retief
WAKKERSTROOM
Wakkerstroom
Standerton
ANDERTON
Charlestown
Lunebery
Lourenço Marques
Delagoa Bay
Catembe
TONGA
LAND
Manhiça
A F R I C A N

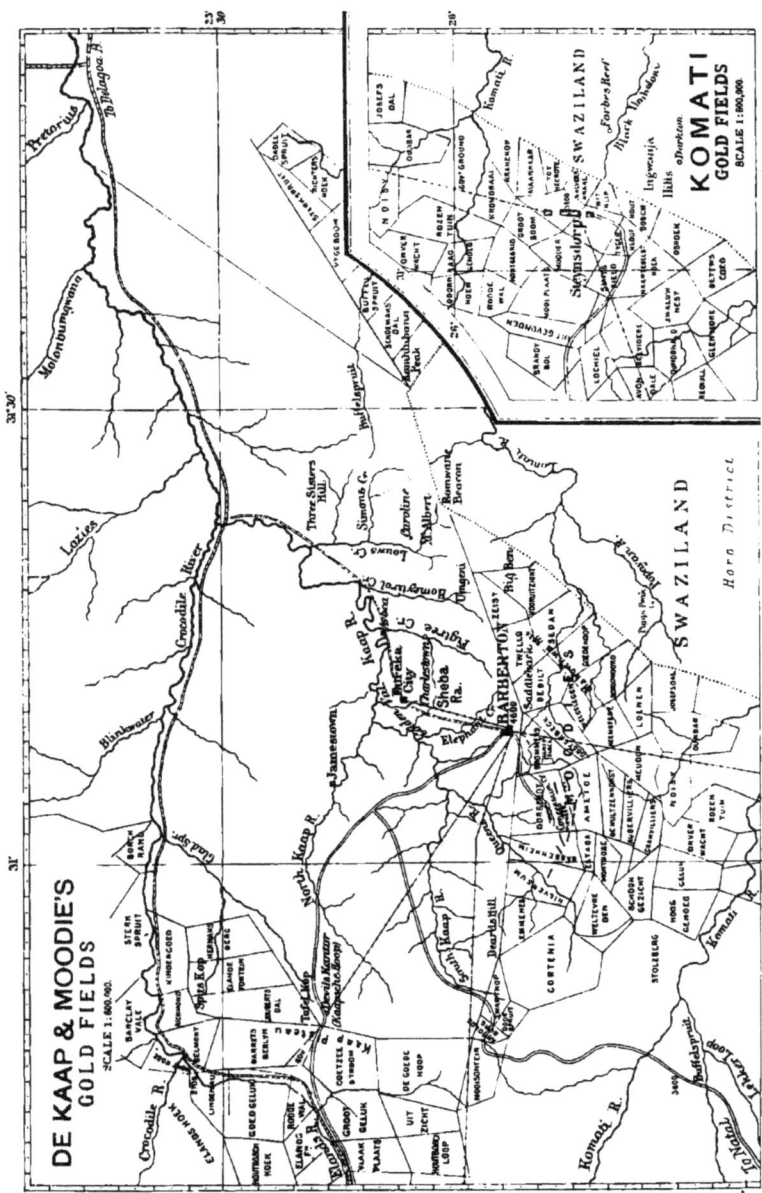

DE KAAP & MOODIE'S
GOLD FIELDS
SCALE 1:600,000.

KOMATI
GOLD FIELDS
SCALE 1:800,000.

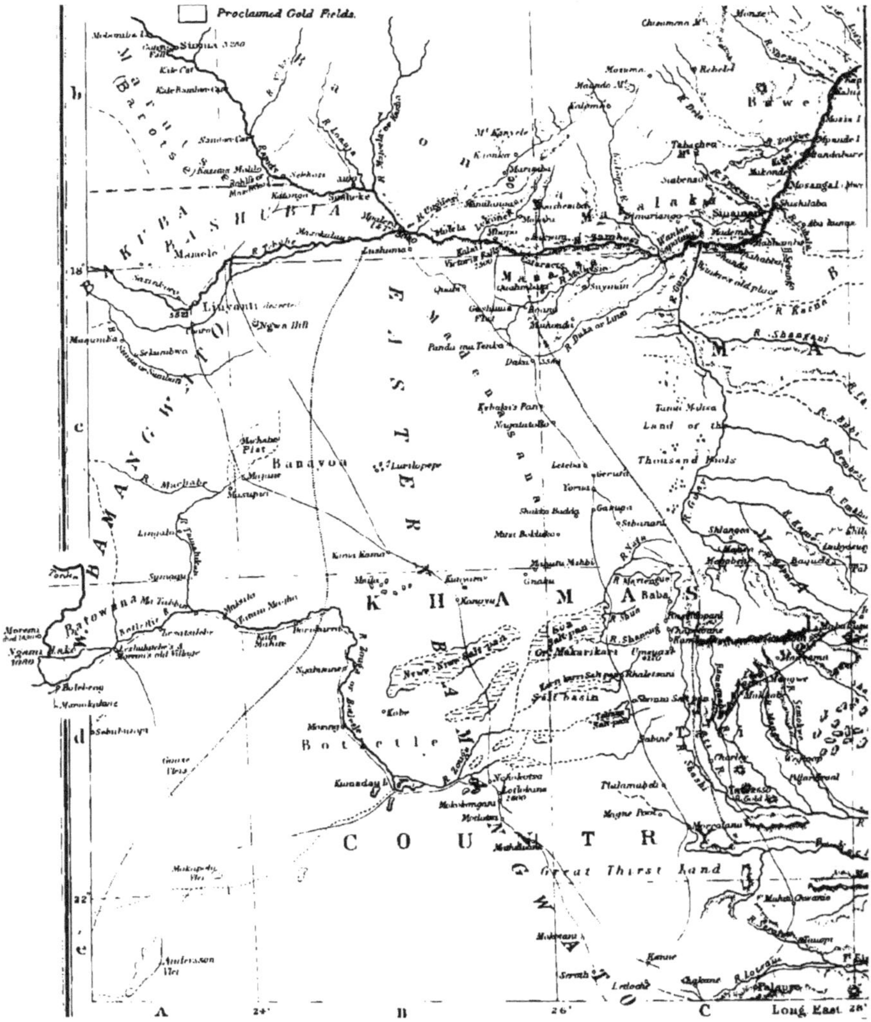
Proclaimed Gold Fields.
Livyanti
Ngwa Hill
Banavoa
Batowana
Ngami Lake
Botletle
COUNTRY
Great Thirst Land
Land of the Thousand Pools
KHAMAS
Victoria Falls
Mababe Flat
R. Mababe
Salt pan
Salt basin
Anderson Vlei
R. Shasha
R. Shangani
R. Gwai
R. Zambesi
Great Salt pan
Tati
Long. East
A
B
C
24°
26°
28°
18°
22°
b
c
d
e

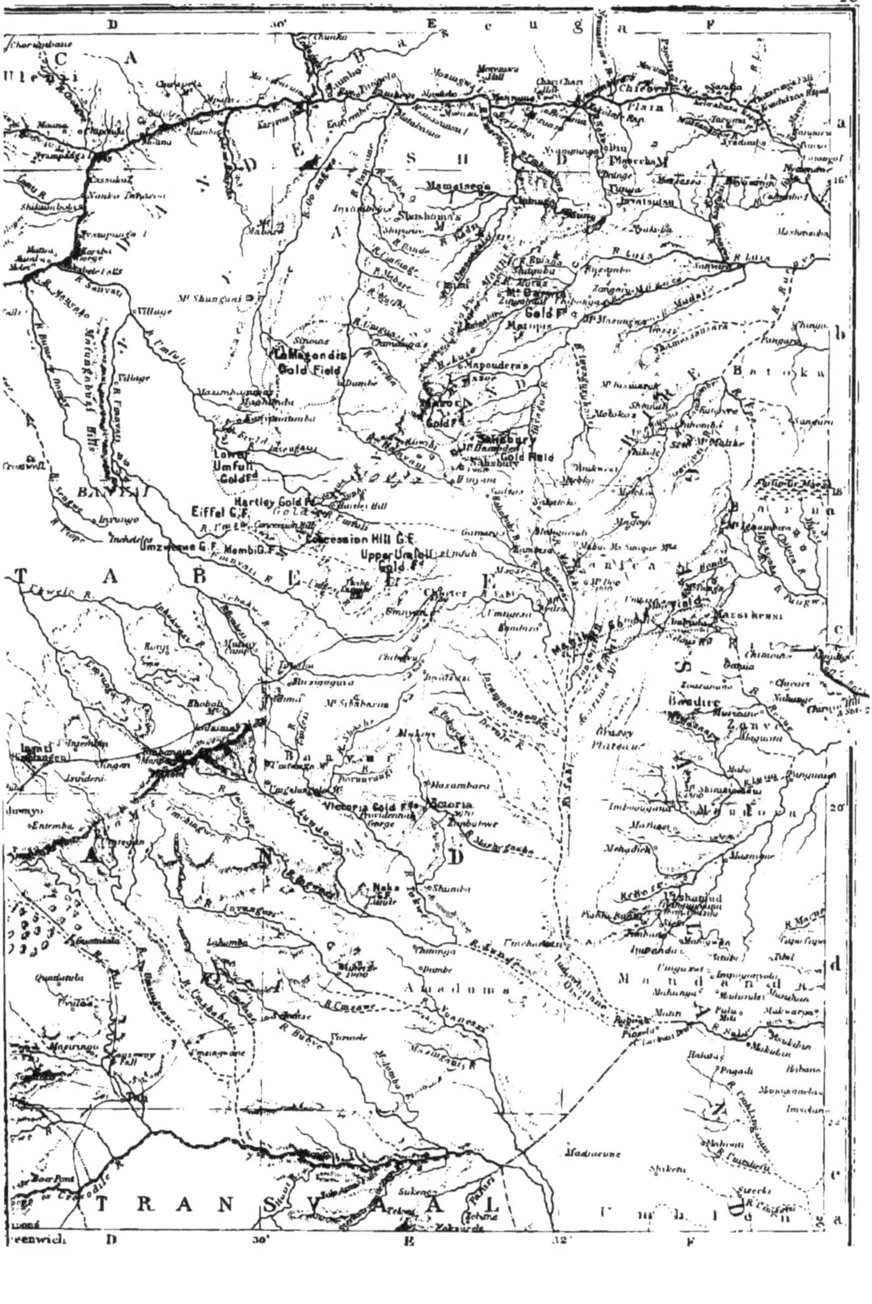
Lo Magondis Gold Field
Mazoe Gold Fd
Salisbury Gold Field
Lower Umfuli Gold Fd
Hartley Gold Fd
Eiffel G.F.
Concession Hill G.F.
Umzweswe G.F.
Upper Umfuli Gold Fd
Victoria Gold Fds
Victoria
Mapoudera's
Matopis
Mamatsega
Chibuga
Chicova
Charter
Manica
Massikessi
Grassy Plateau
Zimbabwe
Amadumu
TRANSVAAL
Greenwich
30°
32°
16°
18°
20°
D
E
F

N O R T H E R N   Z A M B E S I A
R I T I S H   C E N T R A L   A F R I C
Kingdom of Marutse & Mabunda
K I P
Lake Bangweolo or Bemba
Sitanda
Katanga
Lokinga
Marandu
Kabompo
Sesheke
Makalaka
MASHONA LAND
Ft Salisbury
R. Zambesi
R. Chobe
Sapp. Sapp
Kanika's King
Long East 30° of Greenwich
A
B
C
b
c
d
10°
15°
25

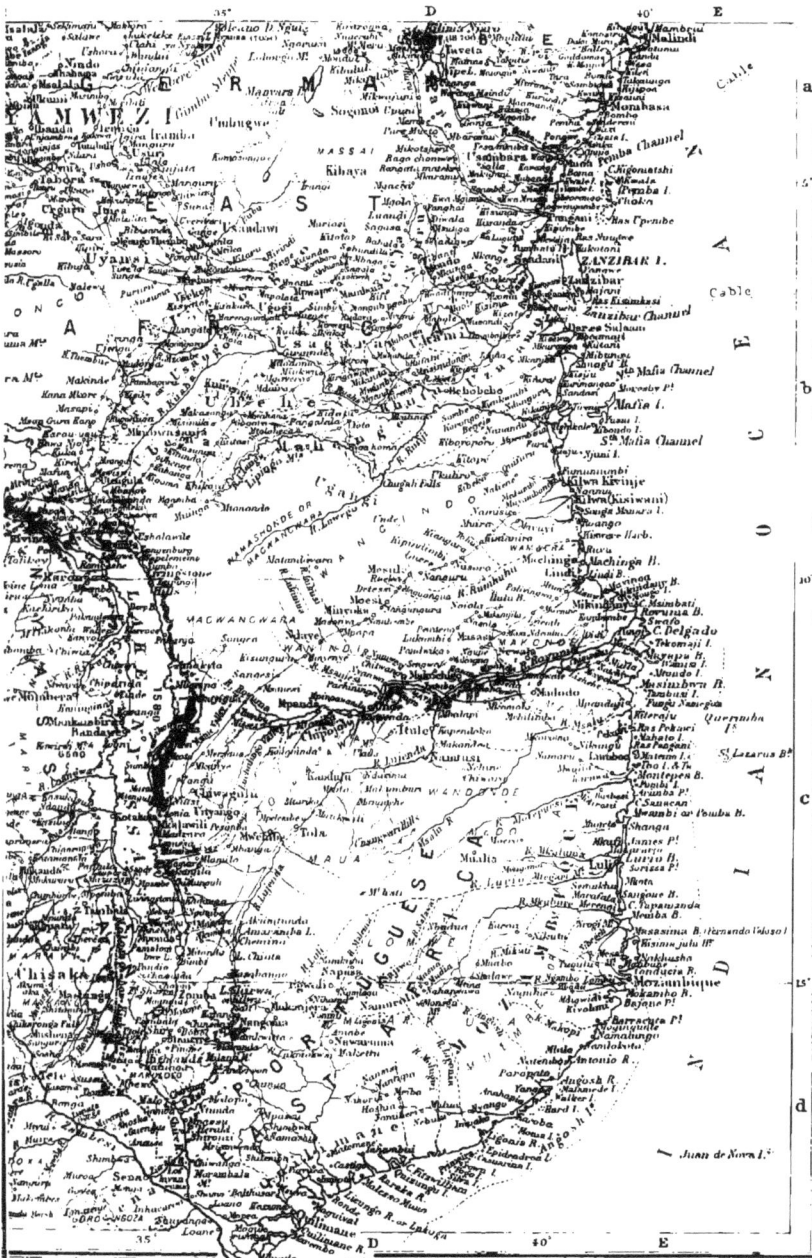

INDEX.

INDEX.

ABBREVIATIONS.

B. = Day.
Bas. = Basutoland.
B.C.A. = British Central Africa.
Bech. = Bechuanaland.
C. = Cape.
C.C. = Cape Colony.
C.F.S. = Congo Free State.
Co. = County.
Dist. = District.
Div. = Division.
E. = East.
Fn. = Fontein.
Ft. = Fort.
Gaz. = Gazaland.
G.E.A. = German East Africa.
G.F. = Gold Field.

Grt. = Great.
G.S.W.A. = German South-West Africa.
Harb. = Harbour.
Hd. = Head.
I. = Island.
Is. = Islands.
Junc. = Junction.
Kr. = Kraal.
L. = Lake.
Lit. = Little.
Mad. = Madagascar.
Mash. = Mashonaland.
Mat. = Matabeleland.
Mt. = Mount or Mountain.
Mts. = Mountains.

N. = North.
Nat. = Natal.
Ny. = Nyasaland.
N.Z. = Northern Zambesia.
O.F.S. = Orange Free State.
P.E.A. = Portuguese East Africa.
Pk. = Peak.
Pks. = Peaks.
Pt. = Point.
P.W.A. = Portuguese West Africa.
R. = River.
Rk. = Rock.
Rks. = Rocks.
S. = South.

S.A. = South Africa.
S.A.R. = South African Republic.
S.E.A. = South-East Africa.
Spr. = Spruit.
Sta. = Station.
S.W. = South-West.
Sw. = Swaziland.
S.W.A. = South-West Africa.
S.Z. = Southern Zambesia.
Tong. = Tongaland.
Tr. = Tribe.
W. = West.
W.A. = West Africa.
Zul. = Zululand.

A

AAPIES R., S.A.R. ... D d 13
Aasvogel Berg, The, C.C. ... F g 8
Aasvogel Pt., C.C. ... D g 8
Aba-hurapi, S.Z. ... C b 15
Abbotsdale, C.C. ... C f 8
Abelakop, The, C.C. ... F c 8
Abercorn, B.C.A. ... E c 3
Aberdeen (Aberdeen), C.C. ... D e 9
Aberdeen (Victoria), C.C. ... F e 9
Aberdeen Road Sta., C.C. ... D c 9
Abiam, G.S.W.A. ... B a 7
Abotak, G.S.W.A. ... A b 7
Abotle, N.Z. ... B c 16
Achte Roggeveld, The, C.C. ... E d 8
Ada, S.A.R. ... F d 13
Adams, Nat. ... D e 16
Adamshoop, C.C. ... E b 7
Addo Heights, C.C. ... E f 9
Adelaide, C.C. ... F e 9
Adendorp, C.C. ... D e 9
Africa, Brit. Cent., C.A. ... D d 3
Africa, German South-West, S.A. ... B f 3
Africa, Portuguese East, E.A. ... D c 16
African Republic, South, S.A. ... B c 12
Agatha, S.A.R. ... F b 13
Agatha, Old, S.A.R. ... F b 13
Aguilhas, Cape, C.C. ... D g 8
Ahilombo, C.F.S. ... A b 10
Ahimbe, Lake, C.F.S. ... B b 10
Aiais, G.S.W.A. ... A b 7
Aikaus, C.C. ... A b 7
Aikhous, G.S.W.A. ... A a 7
Aintas Is., Bech. ... C a 7
Ajawa Tr., P.E.A. ... D c 16
Akananga, C.F.S. ... A c 10
Akuminka, P.E.A. ... C d 16
Akumtunda, P.E.A. ... D e 16
Albany, dist., C.C. ... F f 9
Albasini, S.A.R. ... D b 12
Albert, Nat. ... D d 16
Albert, dist., C.C. ... E d 9
Albert Edward, Fort, ... D c 12
Albert, Fort, S.A.R. ... D c 12
Albert, Fort, Zul. ... E e 16
Albertina, O.F.S. ... C c 10
Albert Silver Mine, S.A.R. ... D d 13
Alcock, S.A.R. ... D d 13
Alexandra, Nat. ... D d 16
Alexandra, S.A.R. ... E b 13
Alexandra, co., Nat. ... D e 16
Alexandria, C.C. ... F f 9
Alexandria, dist., C.C. ... E f 9
Alfred, co., Nat. ... C e 16
Alfred, Fort, C.C. ... F f 9
Algoa Bay, C.C. ... E f 9
Alice, C.C. ... F e 9
Alicedale, C.C. ... E f 9
Aliwal North, C.C. ... F c 10
Aliwal North, dist., C.C. ... E c 10
Aliwal South, C.C. ... A g 9
Allenholm, Nat. ... D e 16
Allison, S.A.R. ... D d 12
All Saints, C.C. ... B f 10
Amabe, P.E.A. ... D d 16

Amadah, G.S.W.A. ... A b 4
Amaqlmua Tr., Mash. ... E d 15
Amagoto Tr., C.C. ... C f 10
Amahlongwa, Nat. ... D e 10
Amajuba Hill, Nat. ... C b 16
Amakita, P.E.A. ... C c 16
Amalienstoin, C.C. ... F f 8
Amandelboom, C.C. ... E d 8
Amazamyama R., Nat. ... G b 7
Amarsinha Lake, P.E.A. ... D c 16
Amas, G.S.W.A. ... B a 7
Amassngo, C.C. ... F c 10
Amasunda, Rand, S.A.R. ... E b 10
Amatikulu, Zul. ... E d 16
Amatola Mts., C.C. ... G e 9
Amatongaland, E.A. ... E g 3
Ambondiro, Mad. ... G f 3
Ambo R., S.Z. ... C b 15
Amersfoort, S.A.R. ... D d 12
Amganul, G.S.W.A. ... B b 7
Amganuros, G.S.W.A. ... A c 7
Amiel, Fort, Nat. ... C b 10
Amina R., C.C. ... A c 9
Amis, C.C. ... A c 7
Amkhous, G.S.W.A. ... B a 7
Amna R., Tk., C.C. ... B b 9
Amos Poort, C.C. ... B e 9
Amos R., C.C. ... B e 9
Ampala, C.F.S. ... B c 16
Amsterdam, S.A.R. ... F e 13
Amutuni, Great, G.S.W.A. ... A a 4
Anwell, S.A.R. ... F a 7
Anahapis, P.E.A. ... D d 13
Aneuse, C.C. ... C d 16
Andara, G.S.W.A. ... C e 3
Anderson Berg, S.A.R. ... F e 13
Anderson, Fort, Ny. ... D d 16
Andersson Vlei, Bech. ... A c 15
Andrade, P.E.A. ... F e 16
Angola, C.F.S. ... A b 16
Angosh Is., P.E.A. ... D d 16
Angosh R., P.E.A. ... D d 16
Angra Pequena, G.S.W.A. ... A g 3
Aninus, C.C. ... A b 7
Anis, G.S.W.A. ... A b 4
Anm's Villa, C.C. ... E f 9
Anoerugas, Bech. ... B c 4
Ansurl, C.F.S. ... B b 16
Antonios Berg, C.C. ... C f 9
Antonio D., P.E.A. ... E f 16
Anys Berg, C.C. ... E f 8
Apies R., S.A.R. ... C c 12
Arakaap R., Tk., C.C. ... B b 8
Arawli Berg, C.C. ... B b 9
Arcona, S.A.R. ... D c 12
Arguin, C.C. ... A b 7
Arirn, Bech. ... B c 4
Ariuela Point, P.E.A. ... E c 16
Aroroams, G.S.W.A. ... A b 7
Aroams, G.S.W.A. ... A c 4
Arundel, C.C. ... E c 9
Assegaai R., Sw. ... E b 10
Atchewa Tr., B.C.A. ... C c 16
Atlantic Ocean, The, S.W.A. ... B d 3
Atys, G.S.W.A. ... A b 7
Auas R., G.S.W.A. ... B a 7
Aub R., G.S.W.A. ... B a 7
Auckland Park, S.A.R. ... 13A
Aughrabis Falls (Orange R.) Great, S.A. ... B a 8

Ankotowa Kraal, C.C. ... A b 7
Annas, G.S.W.A. ... B a 7
Anta Naueji, C.F.S. ... A b 16
Auuns, G.S.W.A. ... A a 4
Avoca, C.C. ... D b 7
Avoca, S.A.R. ... G d 13
Avontuur, C.C. ... C f 9
Ayliff, Fort, C.C. ... G e 9

B

Baba, Bech. ... C d 15
Baba Aijawa Lake, B.C.A. ... B e 16
Babalong, C.C. ... G c 7
Babanango, S.A.R. ... E e 10
Babels Tower, C.C. ... E d 7
Babesi Tr., C.C. ... G e 10
Babbsi Tr., B.C.A. ... E d 3
Babolong, C.C. ... B e 10
Babylon's Tower, C.C. ... D g 8
Bads Berg, S.A.R. ... D e 13
Bablorkwa Tr., Mat. ... D b 4
Bain's Kloof, C.C. ... C f 8
Bajone Point, P.E.A. ... E d 16
Bakalahari, Bech. ... C b 4
Bakenkop, C.C. ... B a 9
Bkkgat, S.A.R. ... F d 13
Bakolok Tr., G.S.W.A. ... B a 4
Bak R., G.S.W.A. ... B a 7
Bakube Tr., G.S.W.A. ... A b 15
Bakwena Tr., Bech. ... A b 13
Bakwena Tr., C.C. ... B c 10
Bakweni Tr., Bech. ... C b 4
Backwini, B.C.A. ... B b 15
Balgowan, Nat. ... D d 10
Balmoral, C.C. ... B f 7
Balmoral, S.A.R. ... D d 13
Balthasar, P.E.A. ... D d 16
Bamagandu, P.W.A. ... A a 4
Bamangwato Tr., East, Bech. ... Bc-Bc 15
Bamangwato Tr., West, Bech. ... A c 15
Bamboes Bergen, The, C.C. ... E d 8
Bamboo Spruit, P.E.A. ... 3A
Bampela Tr., S.A.R. ... C b 12
Banyoza, Bech. ... B c 15
Bandawe, Ny. ... E d 3
Bandini, P.E.A. ... D d 15
Bandire, dist., P.E.A. ... F e 15
Band Spruit, C.C. ... F c 9
Bango, Ny. ... C c 16
Bangkwaketsi Tr., Bech. ... B b 4
Bangweolo, Lake, C.A. ... B c 16
Bankshift, O.F.S. ... E a 9
Bannantyne, S.A.R. ... C c 12
Banyai Tr., S.Z. ... D d 16
Banya Tr., S.Z. ... D d 16
Banyeka Tr., S.Z. ... D b 15
Banye Vlei, C.C. ... B d 8
Baquels, Mat. ... C c 15
Barberton, S.A.R. ... G d 13
Barkly East, C.C. ... A c 10
Barkly Junction, C.C. ... E d 7
Barkly Pass, C.C. ... A f 10
Barkly West, C.C. ... D a 9
Barkly West, dist., C.C. ... C a 9
Baroda, C.C. ... A f 9

Baroka Tr., S.A.R. ... D b 12
Barolong Tr., Bech. ... B c 4
Barotse Tr., B.C.A. ... B c 16
Barraconta, Cape, C.C. ... F g 8
Barracuta Point, P.E.A. ... E d 16
Barren Karroo, The, C.C. ... C c 8
Barroe, C.C. ... D f 9
Barrow, South, Nat. ... D c 10
Barrydale, C.C. ... E f 8
Barno Tr., P.E.A. ... C d 16
Barwari Tr., Bech. ... B c 4
Basaruto I., P.E.A. ... F f 3
Basenga Tr., P.E.A. ... C d 16
Bashee R., C.C. ... B f 10
Bashuis Fontein, C.C. ... B c 7
Bashona Tr., Bech. ... B a 4
Bashubia Tr., G.S.W.A. ... A b 15
Basoetla Tr., S.A.R. ... D b 12
Basson, S.A.R. ... F a 7
Basutoland, S.A. ... A d 10
Bathurst, C.C. ... F f 9
Bathville, S.A.R. ... 13A
Batlaros, Bech. ... B c 4
Batlaro Tr., Bech. ... B c 4
Batlokoa, S.A.R. ... D b 12
Batoka Tr., B.C.A. ... B d 16
Batoka Tr., C.A. ... C d 16
Batonga Tr., B.C.A. ... B b 15
Batowana Tr., Bech. ... A d 15
Batunda, B.C.A. ... A c 16
Baviaanskloof Mts., C.C. ... C f 9
Baviaanskloof R., C.C. ... E e 9
Bavians R., C.C. ... E o 9
Baviens Tr., G.S.W.A. ... B a 4
Bawe, B.C.A. ... C b 15
Bawe Tr., B.C.A. ... B d 16
Bayzeia, C.C. ... B f 10
Bazizia Tr., S.Z. ... B d 16
Bazja, C.C. ... G e 7
Beaconsfield, C.C. ... D a 9
Beaufort, Fort, C.C. ... F e 9
Beaufort, Port, C.C. ... E g 8
Beaufort, West, C.C. ... B e 9
Bechuanaland, S.A. ... B b 4
Bedford, C.C. ... F e 9
Beenbrock, C.C. ... D a 8
Beersoha, O.F.S. ... F c 7
Beerselia, S.A.R. ... C d 13
Beer Vlei, C.C. ... C e 9
Beest Berg, The, C.C. ... C c 3A
Beira, P.E.A. ... 3A
Beira Railway, The, ...
Beira, P.E.A. ... 3A
Belelas Berg, The, S.A.R. ... D b 10
Belgium, New, S.A.R. ... D b 13
Bellmont, C.C. ... E b 7
Bell, C.C. ... G f 9
Bellevue, S.A.R. ... 13A
Bellows Rock, C.C. ... C g 8
Belmont, C.C. ... D b 9
Belvidere, C.C. ... B g 9
Bemba, Lake, C.A. ... B c 16
Bembe, R., P.E.A. ... E b 12
Bembesi, R., Mat. ... C c 15
Bembesville, C.C. ... C f 8
Bensonville, C.C. ... G c 9
Benbrook, C.C. ... A d 10
Bereng, Bas. ... A d 10
Bergendal, S.A.R. ... F d 13
Berg, R., O.F.S. ... D b 9
Berg, R., Great, C.C. ... C e 8
Berlin, C.C. ... G e 9
Bersheba, G.S.W.A. ... B g 3

Entry	Ref	Pg
Bertram, S.A.R.		13A
Beshult Kuil, S.A.R.	D b	12
Bester, O.F.S.	G a	7
Bethanie, O.F.S.	F b	9
Bethanie, S.A.R.	C d	13
Bethany, G.S.W.A.	B g	3
Bethel, C.C.	B e	8
Bethel, S.A.R.	E e	13
Bethelsdorp, C.C.	E f	9
Bethesda, Bas.	A e	10
Bethesda, S.A.R.	E b	13
Bethesda, New, C.C.	D d	9
Bethlehem, O.F.S.	B e	10
Bethulie, O.F.S.	E c	9
Beyers Berg, The, C.C.	B d	9
Bezondermeid, C.C.	A b	7
Bezuidenhout, S.A.R.	G a	7
Bezuidenville, S.A.R.		13A
Biddulphs Berg, O.F.S.	A c	10
Biedouw, The, C.C.	C e	8
Biejespoort, C.C.	B d	9
Bier Spruit, S.A.R.	B d	13
Biggarsberg, The, Nat.	C c	13
Bimbi, Ny.	D e	10
Bird Is., C.C.	F g	9
Bire, Mash.	F b	15
Biribesi, P.E.A.	E c	16
Birthday, S.A.R.	F b	13
Bismiti R., P.E.A.		3A
Bismarck, Mt., Mash.	F b	15
Bisombe, R., B.C.A.	C e	16
Bitter Puits, C.C.	D b	8
Bitter R., C.C.	B e	8
Bitters Fonteins, C.C.	B e	7
Blaashalg Spruit, O.F.S.	F c	9
Blaauwbank, Nat.	C e	10
Blaauw Berg, The, S.A.R.	D b	13
Blaauwbosch Fontein, C.C.	D b	7
Blaauwheuvel, O.F.S.	E c	9
Blaauwkop, The, C.C.	C b	9
Blackburn, Nat.	E d	10
Black Hills, S.A.R.	F b	13
Black Kei R., C.C.	F e	9
Black Umvolosi R., Zul.	E c	10
Blanco, C.C.	B f	9
Blaney Junction, C.C.	G e	9
Blankomo Mts., C.C.	B e	10
Blantyre, Ny.	F e	3
Blauw Bosh Kalk, Bech.	C b	7
Blesbok, S.A.R.	E a	7
Blesbok R.(Heidelberg), S.A.R.	D e	13
Blesbok R.(Standerton), S.A.R.	E e	13
Blignaut's Pont, O.F.S.	E a	9
Blikfontein, C.C.	E a	7
Blinkklip, C.C.	D b	7
Bloed R., S.A.R.	E b	13
Bloemfontein, G.S.W.A.	B a	7
Bloemfontein, O.F.S.	F b	9
Bloemhof, S.A.R.	E a	7
Blood R., C.C.	D f	8
Blood R., S.A.	D d	12
Blood River Sta., C.C.	F f	8
Blue Kop, The, C.C.	C f	9
Blyde R., The, C.C.	C f	9
Blyde R., S.A.R.	F c	13
Blydewerwacht, G.S.W.A.	B b	3
Blytheswood, C.C.	B g	10
Boana, Bech.	B e	15
Boatlanama, Bech.	B b	12
Bobo, Ny.	D d	10
Bobos, S.Z.	F d	15
Bochiapuka, Bech.	B b	12
Bodenstein, S.A.R.	E a	7
Bodiam, C.C.	G f	9
Boer Pont, Bech.	D e	15
Boetsap, C.C.	E a	7
Bohunje, R., N.Z.	B e	16
Bokkeberg, The, C.C.	C d	7
Bokkeveld Bergen, C.C.	D d	8
Bokkeveld, Cold, C.C.	D e	8
Bokkeveld Flats, C.C.	C d	8
Bokkeveld Karroo, C.C.	C d	8
Bokkeveld Mts., Cold, C.C.	D e	8
Bokkeveld, Warm, C.C.	D f	8
Bokkeveld, The, C.C.	B d	8
Bok Point, C.C.	C f	8
Boksburg, S.A.R.	D e	13
Bokstong, Bech.	A d	15
Bolengue Gorge, B.C.A.	D a	15
Boliteletse, Bech.	B e	4
Bolo, C.C.	F d	7
Bolotwa, C.C.	G d	9
Bombai, P.E.A.	B e	13
Bomuningani, Bech.	B g	10
Bomvanaland, C.C.	B g	10
Bomvana Tribe, C.C.	G d	7
Bondlezwarts Territory, G.S.W.A.	B b	3
Bonga, P.E.A.	C d	16
Bonokoe, C.C.	D b	8
Bontekoe, C.C.	D b	8
Booi-en, S.A.R.	D e	12
Boomplaats, C.C.	E b	7
Buroma, P.E.A.	C d	16
Borrels Kopje, C.C.	D a	6
Boschjes Pan, C.C.	C b	9
Boschluis, C.C.	D b	5
Bosch R., C.C.	D e	8
Boshof, C.C.	E b	7
Boshof, O.F.S.	E a	9
Boshof (Heidelberg), S.A.R.	D e	13
Boshof (Watersberg), S.A.R.	C d	12
Bosi R., P.E.A.	F d	15
Bosjes Pan, C.C.	D b	7
Bosworth, C.C.	D d	9
Boterkloof, C.C.	D d	8
Boterlegte, C.C.	F e	8
Botha (Middelburg), S.A.R.	D e	12
Botha (Pretoria), S.A.R.	C c	12
Botha (Zoutpansberg), S.A.R.	C b	12
Botha (Zoutpansberg), S.A.R.	D b	12
Botha Berg, The, S.A.R.	E d	13
Botharnia, S.A.R.	B f	13
Bothasberg, The, O.F.S.	F e	13
Botha's Drift, O.F.S.	E c	7
Botha's Hill, town, C.C.	F f	9
Botha's Pass, S.A.	C b	10
Botlefif, R., Bech.	D d	15
Bot, R., C.C.	D g	8
Botshabelo, S.A.R.	C e	12
Bowan, P.E.A.	D e	13
Bowker, Fort, C.C.	B g	10
Brabis, C.C.	B a	7
Brackenbury, C.C.	B a	9
Brack Pans, Bech.	C a	7
Brack R., S.A.R.	E a	13
Brakfontein (Carnarvon), C.C.	F a	8
Brakfontein (Victoria West), C.C.	C d	9
Brakjes Pan, C.C.	C b	9
Brakpan, S.A.R.	D e	13
Brak, R. (Little Bushman Land), C.C.	C b	8
Brak, R. (Little Nanaland), C.C.	A d	4
Brak, R. (Richmond), C.C.	C e	9
Brak, R. (Somerset East), C.C.	E e	9
Brak, R., S.A.R.	C b	12
Brak, R. (Swellendam), C.C.	E f	8
Brak, R. (Vanrhynsdorp), C.C.	B d	8
Brak, R. (Worcester), C.C.	D f	8
Brak, R., Great, C.C.	F d	9
Brak, R., Little, C.C.	F d	9
Brak Spruit, S.A.R.	A e	13
Brak Valley, C.C.	C d	9
Brandewys ctat, C.C.	B e	9
Brandfort, O.F.S.	F b	7
Brand, R., C.C.	F e	8
Brand Vlei, C.C.	E e	8
Brandvley, S.A.R.	C e	12
Brandwacht, C.C.	E d	8
Braunschweig, C.C.	G e	9
Brazen Head, C.C.	C f	10
Bredasdorp, C.C.	E g	8
Breede, R., C.C.	D e	8
Breede R. Station, C.C.	D f	8
Breidbach, C.C.	G e	9
Breidenbach, S.A.R.	D d	12
Breipaal, O.F.S.	F c	9
Bremersdorp, Sw.	G e	13
Brey Paal, C.C.	C b	7
Briedenhaml, S.A.R.	E a	7
Brink (Rustenburg), S.A.R.	B e	12
Brink (Watersberg), S.A.R.	C c	12
Britannia Reef, C.C.	B e	8
British Central Africa, B.C.A.	B e	16
Broakhorst Spruit, S.A.R.	D d	13
Brown, Fort, C.C.	F f	9
Bruintjes Hoogte, C.C.	E e	9
Bruintjes Hoogte, Klein, C.C.	E f	9
Brul Kolk, C.C.	B b	7
Bua R., Ny.	C e	16
Bubi R., Mat.	C c	15
Bubye R., Mat.	C e	15
Buckingham, Fort, Nat.	D e	10
Buck Kraal, C.C.	F f	9
Buffalo R., G.S.W.A.	B b	3
Buffalo R. (East London), C.C.	A b	10
Buffalo R. (Murraysburg), C.C.	C d	9
Buffel R., C.C.	D d	8
Buffel R., S.A.R.	F e	13
Buffels Rock, C.C.	B e	9
Buffels R. (Namaqualand), C.C.	A b	8
Buffels R. (Sutherland), C.C.	F e	8
Buffels River Mine, C.C.	B b	8
Buila Hills, B.C.A.	C b	15
Bukotabela Rand, Bas.	B c	10
Bulberg, O.F.S.	F b	9
Bulfontein, C.C.	B b	9
Bull Point, C.C.	C e	7
Bull R., C.C.	D e	9
Bultfontein, C C.	D a	9
Bultfontein, O.F.S.	F b	7
Buluwayo, Mat.	D d	15
Bulwer, Nat.	C d	10
Bumbawu, P.E.A.		3A
Bumbi, B.C.A.	C e	16
Bume R., N.Z.	D b	15
Bungane, Tong.	E d	12
Bunkeya, C.F.S.	D d	3
Buntingville, C.C.	B f	10
Buntingville, Old, C.C.	C f	10
Burgers, S.A.R.	D c	12
Burgers, Fort, S.A.R.	F c	13
Burgers Hall, S.A.R.	D e	12
Burghersdorp, C.C.	F e	9
Bushman Land, Great, C.C.	D b	8
Bushman Land, Little, C.C.	C b	8
Bushman R., C.C.	P g	9
Bushman R., Nat.	C d	10
Bushman R. Pass, S.A.	C d	10
Bushman's Kop, O.F.S.	F b	9
Bushman's Nek, Nat.	G b	7
Bushveld, The, S.A.R.	C d	13
Butha Buthe, Bas.	B e	10
Butlers, Nat.	D d	10
Butterworth, C.C.	B g	10
Byrne, Nat.	D d	10

C

Entry	Ref	Pg
Cacadu, C.C.	G d	9
Cahimbe I. (R. Zambesi)	F b	15
Cahimuba, C.F.S.	B e	16
Cajongo I. (R. Zambesi)	F a	15
Cala, C.C.	A f	10
Caledon, C.C.	D g	8
Caledon R., S.A.	B c	10
Caledon R., Great, Bas.	G b	7
Caledon R., Little, O.F.S.	B e	10
Caledon River, dist., O.F.S.	F b	9
Calitzdorp, C.C.	F f	8
Calvinia, C.C.	D d	8
Cambridge, C.C.	A g	9
Cambridge, Fort, S.A.R.	B c	12
Camdeboo Mt., C.C.	C o	9
Camdeboo R., C.C.	D e	9
Cammadagga, C.C.	E f	9
Campbell, C.C.	C a	9
Camperdown, Nat.	D d	10
Cana, Bas.	A d	10
Canarie Fontein, C.C.	A e	7
Cancoco, C.F.S.	A c	16
Camdebot Berg, The, C.C.	D d	7
Cango Berg, The, C.C.	F f	8
Cango Caves, The, C.C.	B f	8
Canpneje, C.F.S.	A b	16
Cananga, C.F.S.	B a	16
Capcon Melemo, C.F.S.	A e	16
Cape Colony, S.A.	C e	7
Cape, dist., C.C.	C g	8
Cape Town (plan of), C.C.		
Cape Cago, P.E.A.	F f	3
Capoco, C.F.S.	A b	16
Carmel, O.F.S.	F e	7
Carnarvon, C.C.	B e	9
Carolina, S.A.R.	E e	13
Casamba, C.F.S.	A b	16
Casova, P.E.A.	C e	16
Cassa, C.F.S.	A c	16
Cassuko I. (R. Zambesi)	D b	15
Castigo, P.E.A.	D d	16
Castilhopolis, S.A.R.	E e	13
Castle, Cape, C.C.	B e	8
Casuarina I., P.E.A.	D d	16
Cathcart, C.C.	G e	9
Cathkin Peak, Nat.	C d	10
Cedar Berg, The, C.C.	D d	7
Cedar Mts., C.C.	D e	8
Cedarville, C.C.	B e	10
Centocow, Nat.	C c	10
Central Africa, British, B.C.A.	B e	16
Central Karroo, The, C.C.	C f	9
Ceres, C.C.	D f	8
Ceres Road Sta., C.C.	D f	8
Chabuela, Ny.	C e	16
Chagos, C.C.	E b	7
Chakane, Bech.	C e	15
Chalumna R., C.C.	G f	9
Chama, B.C.A.	B b	16
Chana, C.F.S.	B e	16
Chambezi R., B.C.A.	E d	3
Chamob, W., G.S.W.A.	A c	4
Champagne Castle, S.A.	C b	7
Cham, W., G.S.W.A.	A c	4
Changwazi Hills, P.E.A.	D e	16
Chacoxi, Ny.	D d	16
Chapotoane, Bech.	C d	15
Chari Chari Hill, P.E.A.	E a	15
Charlestown, Nat.	E c	3
Charlestown, Nat.	C b	10
Charlestown, S.A.R.	C d	13
Charley, Tati	C d	15
Charlton, C.C.	E d	9
Charo, Bech.	A e	15
Charter, Mash.	E c	15
Chasa, B.C.A.	C e	16
Chasaya, P.E.A.	C c	16
Chasumba, Ny.	C c	16
Chater, Fort, Zul.	E c	10
Chelmsford, Fort, Zul.	E c	10
Chelsea Point, C.C.	E g	9
Chilanda, B.C.A.	C e	16
Chilneyu, Mash.	C e	16
Chilinga, P.E.A.	C d	16
Chitongata, Mash.	F b	15
Chibuln, C.F.S.	A b	16
Chibwe, B.C.A.	C b	16
Chicari, P.E.A.	F c	15
Chicombo, Mash.	C d	16
Chicova, P.E.A.	F a	15
Chicova Plain, P.E.A.	F a	15
Chicualla Cualla, P.E.A.	G a	13
Chicundo, S.A.R.	D b	4
Chigaragara, Ny.	C c	16
Chilia R., S.A.R.	G a	13
Chilomba, Mash.	F b	15
Chikaronga Fali (R. Zambesi)	C d	16
Chikole, Mash.	F b	15
Chikonta, M., Ny.	C c	16
Chikosi, B.C.A.	B b	16
Chikundu, B.C.A.	C c	16
Chileo, C.F.S.	A b	16
Chiloane, P.E.A.	F f	3
Chinan, P.E.A.	D d	16
Chimanga's, Mash.	E b	15
Chimbimbe, Ny.	C c	16
Chimbimbe, Ny.	F c	15
Chinaaka, P.E.A.	D c	16
Chinanga, P.E.A.	D e	16
Chinanga, B.C.A.	C e	16
Chinde, P.E.A.	F c	3
Chingwayo, S.A.R.	D d	12
Chinoni, Mt., P.E.A.		3A
Chinsuni, Ny.	D d	16
Chipatula, B.C.A.	B b	16
Chipatula, Ny.	C e	16
Chipojola, P.E.A.	D e	16
Chiponga, B.C.A.	D a	15
Chirapela, Mt.	B d	16
Chiromo, Ny.	F e	3
Chirove, B.C.A.	C e	16
Chiruvu Hill, P.E.A.	F e	15
Chiruvu Sta., P.E.A.		3A
Chisaka, dist., P.E.A.	C c	16
Chisamena Mt., B.C.A.	B d	16
Chisunguli, Ny.	C c	16
Chitanga, Mat.	E d	15
Chitembo, B.C.A.	E d	3
Chitesi, P.E.A.	C c	16
Chitimba, B.C.A.	C b	16
Chitokoka, P.E.A.	C c	16
Chitora R., P.E.A.	F c	15
Chitumbe, Ny.	C c	16
Chiwaguba, P.E.A.	D c	16
Chiwanga, P.E.A.	D d	16
Chiwara, P.E.A.	D c	16
Chiwiu, Ny.	C c	16
Chiwiyi, Ny.	C c	16
Chobe, R., S.W.A.	B a	4
Choelitu, G.S.W.A.	A b	4
Chongue, R., B.C.A.	B d	16
Chopo, Bech.	D a	7
Chorumane, B.C.A.	B d	16
Chorumbano, B.C.A.	B d	16
Chosi, R., Ny.	C b	16
Chozi, B.C.A.	C b	16
Christel, Lake, S.A.R.	E a	7
Christian, S.A.R.	E a	7
Christiania Bay, C.C.	B e	7
Chuaka, P.E.A.	E a	15
Chuane Pits, Bech.	B b	4
Chumbi, P.E.A.	C c	16
Chunes R., S.A.R.	C e	10
Chunga, B.C.A.	C b	16
Church Berg, Bech.	B e	4
Chuzu's, Mash.	C d	16
Chwapong, S.A.R.	C b	12
Cio, C.F.S.	A c	16
Clanwilliam, C.C.	C f	8
Claremont, C.C.	C g	8
Clarkebury, C.C.	B f	10
Clarkson, C.C.	D g	9
Clearwater, C.C.	A c	9
Cliff Pt. (Little Namaqualand), C.C.	A c	4
Cliff Pt. (Vaanrhynsdorp), C.C.	C d	8
Clifton (Bedford), C.C.	F e	9

Name		
Clifton (Fort Beaufort), C.C.	F e	9
Cloetes Tafel, The, C.C.	E e	8
Clumber, C.C.	F f	9
Clydesdale, C.C.	C e	10
Cobus Louws R., C.C.	F e	8
Cockscomb Mt., C.C.	D f	9
Coega R., C.C.	E f	9
Coelzee (Bloemhof), S.A.R.	E a	7
Coetzee (Lydenburg), S.A.R.	D e	12
Coernay R., C.C.	E e	9
Coetze, S.A.R.	D d	13
Cogman's Kloof, C.C.	E f	8
Colatto, Cape, P.E.A.	D c	4
Colchester, C.C.	E f	9
Cold Bokkeveld, The, C.C.	D e	8
Coldstream, Nat.	C b	10
Coldstream, S.A.R.	E f	13
Colenso, Nat.	C c	10
Colesberg, C.C.	E c	10
Colosa, C.C.	B g	10
Combrink, C.C.	E b	7
Commadagga, C.C.	E d	7
Commando Drift, O.F.S.	F a	7
Commando, R., O.F.S.	C b	10
Commissioners Salt Pan, C.C.	E e	8
Comoro I., G.A.	G d	3
Compass Berg, The, C.C.	D d	9
Concession Hill, Mash.	E c	15
Concordia Mine, C.C.	B b	8
Condacia Bay, P.E.A.	E e	16
Cone Point, Nat.	F c	10
Conference Hill, S.A.R.	D b	10
Confunvaba, C.C.	G e	9
Constable, C.C.	E f	8
Content, C.C.	D a	9
Conway, C.C.	E d	9
Cookhouse, C.C.	E e	9
Coopersdal, S.A.R.	D e	12
Cornelis, R., O.F.S.	C b	10
Corrientes, Cape, P.E.A.	F f	3
Cove Rock, C.C.	A b	10
Covie, C.C.	C f	9
Cradock, C.C.	E c	9
Cradock Fontein, C.C.	E f	9
Cradock, Fort, Zul.	E d	10
Crocodile R., S.A.	D b	9
Crocodile R., S.A.R.	F d	13
Crocodile R., S.Z.	E d	12
Cronje, S.A.R.	F a	7
Cross, Cape, G.S.W.A.	A f	3
Crown I., P.E.A.	D d	16
Cuanlo, R., P.W.A.	B a	4
Cunningham, C.C.	B g	10
Cunyana, P.E.A.	D b	4
Curtis, Mt., C.C.	C e	10
Curtis, Fort, Zul.	E c	10
Cutagandas, G.F.S.	A e	10
Cyphergat, C.C.	F d	9
Cypress Grove, C.C.	E d	9

D

Name		
DABEGABIS, G.S.W.A.	B b	3
Daberas, G.S.W.A.	A c	4
Dabe, The, C.C.	G a	8
Dadelfontein, S.A.R.	F b	13
Daline, C.C.	G e	9
Dainus, G.S.W.A.	B b	7
Dainge, P.E.A.	F b	15
Daka, Bech.	B c	15
Daka, R., Bech.	C c	15
Dakaraland, G.S.W.A.	A c	4
Dambe, Mash.	E b	15
Dande R., S.Z.	C b	15
Danger Point, C.C.	D g	8
Daniels Kuil, C.C.	C a	9
Dannhauser, Nat.	D b	10
Darde Bergen, The, S.A.R	E a	13
Dargle Road Sta., Nat.	D d	10
Darika, P.W.A.	B a	4
Darkton, Sw.	G e	13
Darling, C.C.	E b	8
Darwin G. F., Mt., Mash.	E b	15
Dassen, C.C.	C f	8
Dassenberg, C.C.	B d	7
Dassen R., C.C.	B f	8
Davidsgraf, O.F.S.	D b	9
Dawignab, G.S.W.A.	B a	9
De Aar Junction, C.C.	C c	9
De Beer, O.F.S.	D c	13
De Beers, C.C.	F b	7
De Beers, O.F.S.	E b	7
De Beers Vlei, C.C.	C f	9
Debing, Bech.	D a	4
Debogankia, S.A.R.	B c	12
Dubra, G.S.W.A.	A b	4
De Goupli, C.C.	B d	4
De Jager, S.A.R.	G a	7
De Kaap G. F., S.A.R.	F d	13
De Kloof, C.C.	E b	8
De Kruis (Carnarvon), C.C.	F b	8

Name		
De Kruis (Fraserburg), C.C.	E d	8
Delagoa Bay, P.E.A.	E g	3
Dela, R., B.C.A.	B d	16
Delgado, Cape, P.E.A.	G d	3
Delportsberge, C.C.	D a	9
Denikane, S.Z.	B a	4
De Pat, O.F.S.	E b	9
Dephiriog, Bas.	F b	7
Derby, S.A.R.	F e	13
Derde Poort, S.A.R.	B c	12
De Riet, C.C.	A c	7
Descania, Cape, C.C.	C e	8
De Tuin, C.C.	C b	7
Deuka, P.E.A.	C c	16
Devil's Kantor, S.A.R.	F d	13
Devule R., Mosh.	E c	15
Dewetsdorp, O.F.S.	F b	9
Didema, Mt., C.C.	F e	9
Diko's, C.C.	C e	10
Dilolo, Lake, B.C.A.	C d	3
Dinah, Fort, C.F.S.	B b	10
Dinizulu Tribe, Zul.	E c	10
Dipetnng Nek, Bas.	B d	10
Dipnino, C.F.S.	A b	16
Disselsdorp, C.C.	B f	9
Diu, P.E.A.	C d	16
Djelele, R., S.A.R.	F a	13
Doe Mt., S.A.R.	F e	13
Dombe Berg, The, S.A.R.	D b	16
Domodhlomo, S.A.R.	E b	10
Donoe Mt., P.E.A.	C d	16
Donald, Fort, C.C.	C e	10
Donkin Bay, C.C.	C d	8
Doorn Bergen, The, C.C.	B b	9
Doornbosch, C.C.	D e	8
Doornfontein, Bech.	D a	7
Doornkop, The, S.A.R.	B d	13
Doorn R. (Clanwilliam), C.C.	C d	8
Doorn R. (Jansenville), C.C.	D f	9
Doorn R., Karroo, C.C.	C d	8
Doorn R., Zwart, C.C.	B c	8
Doorns, C.C.	D b	7
Doorn Spruit, C.C.	E b	7
Dora, Lake, P.E.A.		3a
Dordrecht, C.C.	F d	9
Dorokarra, Bech.	D d	15
Dorps R., S.A.R.	E b	13
Dornmyamgi R., Mash.	E c	15
Dosefu, C.F.S.	A c	16
Double Mts., Bas.	A d	10
Douglas, C.C.	C b	9
Dover, C.C.	E b	7
Draal Fontein, C.C.	C d	7
Drakensberg, The, S.A.	G c	10
Drakensberg, The, S.A.R.	D e	12
Drakenstein Mts., C.C.	E f	8
Dreman Sta., C.C.	E e	9
Dreyer, S.A.R.	E e	12
Driefontein (Boshof), O.F.S.	E a	9
Driefontein (Kroonstad), O.F.S.	A b	10
Dronfield, C.C.	C c	12
Dronkfontein, S.A.R.	C e	12
Droogdep R., C.C.	B b	8
Drooge R., C.C.	C d	8
Drooge Strand, C.C.	C b	7
Drui Fontein, C.C.	E a	10
Duiker Point, C.C.	G g	8
Duivenhoeks R., C.C.	F g	8
Dumbe, Mat.	E d	15
Dunniny Point, C.C.	B e	8
Dundee, Nat.	D c	10
Dundees, O.F.S.	D f	12
Du Plessis, S.A.R.	C c	12
Duplooi, S.A.R.	C e	12
Dupree, S.A.R.	G a	7
Du Pree's, S.A.R.	B d	12
Durban, Nat.	E d	10
Durbun, co., Nat.	D d	10
Durbanville, C.C.	C f	8
Durnford Bay, Zul.	D d	10
Durnford Point, Zul.	F e	10
Dutoit, S.A.R.	D d	12
Dutoits Pan, C.C.	D a	9
Duvenage, S.A.R.	C f	12
Dwaalfontein (Hanover), C.C.	D d	9
Dwaal Fontein (Prince Albert), C.C.	B e	9
Dwars Berge, S.A.R.	B c	13
Dwars R., S.A.R.	B b	13
Dwekwa R., C.C.	D e	8
Dwyka R., C.C.	B f	8
Dyer I., C.C.	D g	8

E

Name		
EAST AFRICA, PORTU-GUESE	D d	16
East London, C.C.	A b	10
Ebenezer, C.C.	C d	8

Name		
Ebenezer, Nat.	H c	7
Ebenezer (Rustenburg), S.A.R.	C c	12
Ebenezer (Wakkerstroom), S.A.R.	D d	12
Edenburg, O.F.S.	E b	9
Edendale, Nat.	D d	10
Eendoorn, C.C.	B c	8
Eersteling, S.A.R.	C c	12
Eerste Poort, S.A.R.	E c	12
Eerste R. Junction, C.C.	C g	8
Egwoli, P.E.A.	D d	16
Ehlanzeni, Nat.	D c	10
Ehluhlangeni, S.A.R.	D d	12
Eiffel Gold Field, Mash.	D c	15
Eljas, C.C.	B b	7
Ekamba, Nat.	B e	10
Ekuhangeni, S.A.R.	E b	10
Elandsberg, The, C.C.	F e	9
Elands Berg, The, S.A.R.	D b	10
Elands Drift, C.C.	D e	8
Elands Fontein, C.C.	C e	9
Elandsfontein, S.A.R.	D e	13
Elandsfontein June., S.A.R.	D e	13
Elands Kloof, C.C.	F f	8
Elandskop, The, O.F.S.	B b	10
Elands R. (Lydenburg), S.A.R.	F d	13
Elands R. (Pretoria), S.A.R.	D d	13
Elands R. (Rustenburg), S.A.R.		
Elands R. (Tarkastad), C.C.	F d	9
Elands R. (Utenhage), C.C.	E f	9
Elebe, Fort, Bech.	C c	15
Elephant R., P.E.A.	B b	13
Elephant Rock, C.C.	C d	8
Elephant Vley, G.S.W.A.	A b	4
Eliss, C.C.	D g	8
Elim, Nat.	D e	10
Elizabeth, Port, C.C.	E f	9
Ellerton, S.A.R.	C d	12
Elliottdale, C.C.	B g	10
Eloff Net, S.A.R.	E c	12
Elsburg, S.A.R.	D c	13
Elukuweni, C.C.	B e	10
Embekelweni, Sw.	G c	13
Embamoneli R., S.A.R.	G d	12
Emfuleni, Zul.	D e	10
Emfundisweni, C.C.	C e	10
Emigratie, S.A.R.	D d	12
Emkanduli, C.C.	B f	10
Empangwene, Zul.	E e	10
Emperor William's Gold Field, Mash.	F b	15
Emmaus, Nat.	C c	10
Emmaus, O.F.S.	D e	12
Encobo, C.C.	B f	10
Endyane, Mat.	C a	4
Engelbrecht (Potchefstroom), S.A.R.	B d	12
Engelbrecht (Pretoria), S.A.R.	C c	12
Engels Berg, The, S.A.R.	B d	13
English Drift, C.C.	D b	7
English R., P.E.A.	E e	12
Enjanyana, C.C.	B f	10
En Kokerboom, C.C.	B c	7
Enon, C.C.	E c	9
Entemba, Mat.	D d	15
Entonjaneni, dist., Zul.	E c	10
Entumeni, Zul.	E c	10
Epedendrom R., Zul.	D b	10
Erasmus (Boshof), O.F.S.	E b	9
Erasmus (Heilbron), O.F.S.	D f	13
Erasmus (Pretoria), S.A.R.	D d	13
Erasmus (Wakkerstroom), S.A.R.	G a	7
Ermelo, Kort, S.A.R.	D d	13
Ermelo, S.A.R.	F e	13
Eshowe, Zul.	E c	10
Estcourt, C.C.	C d	10
Esterhuise, S.A.R.	C d	12
Eugenie, Zul.	E c	10
Eureka City, S.A.R.	G d	13
Evansdale, Nat.	D e	10
Evelyn, Fort, C.C.	B e	10
Ezel Berg, The, C.C.	C b	9
Ezels Fontein, C.C.	E c	7

F

Name		
FAIGONI, Tong.	D c	4
Fairfield, C.C.	F g	8
False, C.C.	C g	8
False Bay, Zul.	F b	10
Fannings Mine, C.C.	B a	8
Faraday, C.C.	B b	9
Fauresmith, O.F.S.	E b	9

Name		
Fernando Veloso Bay, P.E.A.	E c	16
Fern Hill Sta., Nat.	D d	10
Ferreira, S.A.R.		13a
Ficksburg, O.F.S.	A c	10
Fife, B.C.A.	E c	3
Finga, Ny.	C b	16
Fingoes, The, C.C.	F d	7
Fingo Tribe, C.C.	A g	10
Fishasi, C.C.	A g	9
Fisbla Kanbu, P.E.A.	F d	15
Fish Bay, C.C.	A g	9
Fish Point, C.C.	A d	7
Fish R., C.C.	E d	8
Fish R., Great, C.C.	E c	9
Fish R., Little, C.C.	E e	9
Fish River Sta., C.C.	E d	9
Fitzwilliam, Cape, P.E.A.	D d	16
Flat Point, C.C.	B g	10
Flesh Bay, C.C.	A g	9
Fletcher, Fort, C.C.	B e	10
Florence Bay, Ny.	C c	16
Florey, S.A.R.	D c	12
Fogo I., P.E.A.	D d	16
Fokoti, Tong.	F b	10
Fontesville, P.E.A.		3a
Forbes Reef, S.W.	G e	13
Fordsburg, S.A.R.		13a
Fordyce, Fort, C.C.	F e	9
Forest Hall, C.C.	C f	9
Fort Abercorn, B.C.A.	E c	3
Fort Albert, Zul.	E c	10
Fort Amiel, Nat.	C b	10
Fort Anderson, Ny.	D d	16
Fort Ayliff, C.C.	A e	10
Fort Beaufort, C.C.	F e	9
Fort Bowker, C.C.	B g	10
Fort Brown, C.C.	F f	9
Fort Buckingham, Nat.	D c	10
Fort Cambridge, C.C.	B e	10
Fort Charter, Mash.	E c	15
Fort Charter, Zul.	E c	10
Fort Chelmsford, Zul.	E c	10
Fort Cradock, Zul.	E d	10
Fort Curtis, Zul.	C b	10
Fort Dinah, C.F.S.	B b	10
Fort Donald, C.C.	C e	10
Fort Evelyn, S.A.R.	E c	10
Fort Fletcher, C.C.	B e	10
Fort Fordyce, C.C.	F e	9
Fort George, S.A.R.	D e	12
Fort Maguire, Ny.	D c	16
Fort Marshall, S.A.R.	E c	10
Fort Hardy, Bas.	F e	7
Fort Harrison, C.C.	C f	9
Fort Hartley, Bas.	A e	10
Fort Jackson, C.C.	G e	9
Fort Johnston, Ny.	C c	16
Fort Lister, Ny.	D c	16
Fort Pine, Nat.	D c	10
Fort Salisbury, Mash.	E b	15
Fort Sharpe, Ny.	C d	16
Fort Tenedos, Zul.	E d	10
Fort Tuli, Mat.	D d	15
Fort Victoria, Mat.	E c	15
Fort Victoria, S.A.R.	E c	10
Fort Vincent, S.A.R.	D c	10
Fort Warden, C.C.	B g	10
Fort Warwick, S.A.R.	D c	10
Fort William, C.C.	B e	10
Fort William, S.A.R.	C c	12
Fort Wood, Zul.	D f	10
Foulrub R., C.C.	B a	9
Fourie, C.C.	D e	9
Fourie (Rustenburg), S.A.R.	E c	12
Fourie (Wakkerstroom), S.A.R.	D d	12
Fourie, S.A.R.	D c	12
Fouriesburg, O.F.S.	B c	10
Fourteen Streams, C.C.	D a	9
Francks Spruit, S.A.R.	B c	12
Frankfort, C.C.	A g	10
Frankfort, O.F.S.	C e	12
Fraserburg, C.C.	F d	8
Fraserburg Road Sta., C.C.	F e	8
Freemanstown, C.C.	G d	9
French Hock, C.C.	D f	8
Frere, C.C.	C e	10
Frio, Cape, G.S.W.A.	A d	3
Fukuma Mt., B.C.A.	B c	16
Funbi I., P.E.A.	D d	16
Fingafunga, B.C.A.	B b	16
Fungu Namegua I., P.E.A.	D d	16
Funk Tr., Bech.	B b	4
Furumana, B.C.A.	C c	16

G

GABIS, G.S.W.A.	B b	7
Gadaos Ford (Orange R.), S.A.	A a	8
Gaibes, G.S.W.A.	A b	7
Gaikas, The, C.C.	B g	10
Galekas, The, C.C.	B g	10
Gamba, P.E.A.	C d	16
Gameno's, Mash.	E e	15
Gamka R., C.C.	B f	9
Gamogara, Bech.	D a	7
Gantoos R., C.C.	C f	9
Gamzel Vlei, C.C.	A d	9
Ganab, G.S.W.A.	A n	7
Ganda, P.E.A.	F e	15
Gandura Vley, P.E.A.		3 a
Ganesa, Bech.	E a	7
Ganikobis, G.S.W.A.	A c	4
Gans, G.S.W.A.	B b	7
Ganzel Vlei, C.C.	F d	8
Gaozi, Zul.	D e	4
Garel Graf, C.C.	D c	7
Garenganze Tr., C.F.S.	B c	16
Garieb R., Ger., S.A.	C b	9
Gariepine Walls, S.W.A.	D a	8
Garis, C.C.	B c	8
Garis, G.S.W.A.	A b	4
Garou, S.A.R.	C d	13
Garuga, Mat.	C c	8
Gashuma Flat, Bech.	D c	15
Gasip, C.C.	D b	7
Gaspan, S.A.	D b	9
Gatberg, The, C.C.	B f	10
Gats Rand, The, S.A.R.	C c	13
Gavereni, R., Mash.	F b	15
Gaza Land, P.E.A.	F c	16
Gealeka Tr., C.C.	B g	10
Geelbecksvlei, S.A.R.	C e	13
Geelhoutkop, St.,S.A.R.	D c	13
Gelab, G.S.W.A.	B a	7
Geldaos, G.S.W.A.	A b	4
Geigaob, G.S.W.A.	A a	7
Gel Garieb, R., S.A.	C b	9
Geikhaus Tr., G.S.W.A.	A b	4
Gels, G.S.W.A.	B a	7
Geitsaub, G.S.W.A.	B a	7
Geluk Tr., S.A.R.	E c	13
Gemfonkberg, The, C.C.	A b	7
Gemsbok R., C.C.	D c	8
Genadenthal, C.C.	D g	8
George, C.C.	B f	9
George, Fort, S.A.R.	D e	13
Georgengholtz, S.A.R.	F a	13
Gerlekz Point, C.C.	B g	9
Gerlach's Hope, S.A.R.	D c	12
German South-West Africa		
Germiston, S.A.R.	D c	13
Gerufa, Mat.	C e	15
Gert Leow, C.C.	E b	7
Gethsemane, Bas.	A d	10
Gey, S.A.R.	F a	7
Gham, C.C.	C a	8
Ghangkbuara, Bech.	D b	12
Ghanze, Bech.	B b	4
Ghatwani, Bech.	E a	7
Giant's Castle, The, S.A.	C d	10
Gilboa, G.S.W.A.	B g	3
Gilboa, S.A.R.	C d	9
Gift Bergen, The, C.C.	E a	8
Gindundu, P.E.A.	C c	10
Glasgow, New, Nat.	E d	10
Glassen Point, C.C.	E g	9
Glencoe, Fort, Nat.	E e	8
Glencoe Junction, Nat.	D c	10
Glen Connor, C.C.	E f	9
Glendale, Nat.	D d	10
Glenlynden, C.C.	F e	9
Gnadlakka Point, C.C.	E g	9
Gnaku, Bech.	B d	15
Gongtbgaos, G.S.W.A.	A a	7
Goanus, G.S.W.A.	A c	4
Gobabis, G.S.W.A.	B f	3
Gobas, G.S.W.A.	B a	7
Gobatei, Bech.	D a	7
Goedeverwacht, C.C.	C e	8
Goedgedach, S.A.R.	F c	13
Goemens Berg, O.F.S.	D b	9
Gold R., S.A.R.	D a	7
Gomoperi, Bech.	D a	7
Gonamolopue Rand, B.C.A.	D b	15
Gonga, P.W.A.	A a	4
Gong Gong, C.C.	D c	7
Gonin, S.A.R.	B c	13
Gonubie R., C.C.	B g	10
Gonye Falls, B.C.A.	A d	16
Good Hope, Cape of, P.E.A.	C g	8
Goose Vleis, Bech.	A d	15
Gopani, S.A.R.	B c	13
Gordon Bay, C.C.	C f	15
Gordonia, Bech.	F a	5
Gordon's Bay, C.C.	F a	8
Gorima Mts., Mash.	F e	15

Gorongoza Tr., P.E.A.	C d	16
Goschen, C.C.	E e	9
Gosa, Mash.	E b	15
Goulux, C.C.	F d	7
Goubes, C.C.	B e	7
Goub R., G.S.W.A.	B n	7
Goudini, C.C.	D f	8
Gougouzaas, G.S.W.A.	A n	7
Gouwitz R., C.C.	A g	9
Gous Pan, C.C.	D b	7
Govea, P.E.A.	C d	16
Graaff Reinet, C.C.	B e	9
Grahamstown, C.C.	F g	9
Great Aughrabis Falls (Orange R.)	E a	8
Great Berg R., C.C.	C e	8
Great Brak R., C.C.	F d	8
Great Brak River, town, C.C.	B f	9
Great Brak Spruit, S.A.R.	B d	13
Great Bushman Land, C.C.	D b	8
Great Fish R., C.C.	E d	9
Great Fish R., G.S.W.A.	A c	4
Great Kei R., C.C.	B g	10
Great Lion R., C.C.	B f	9
Great Marsh, The, S.Z.	B a	4
Great Oliphants R., S.A.R.	C d	12
Great Palala R., S.A.R.	C b	12
Great Pan, The, C.C.	C a	9
Great Paternoster Point, C.C.	B e	8
Great Riet, C.C.	C b	7
Great Riet R. (Somerset East), C.C.	E e	9
Great Riet R. (Sutherland), C.C.	C c	8
Great Thirst Land, Bech.	C d	15
Great Winterberg, C.C.	F e	9
Great Winter Hoek, C.C.	C f	8
Great Zwart Berg, C.C.	C f	8
Great Zwarte Bergen, The, C.C.	A f	9
Greendoorns R., S.A.R.	F d	9
Greylingstad, S.A.R.	D e	13
Greytown, Nat.	D d	10
Greytown (Robertson), C.C.	D f	8
Greytown (Stutterheim), C.C.	B c	8
Grieve, S.A.R.	F b	13
Griqualand East, C.C.	B e	10
Griqualand West, C.C.	C a	9
Griquatown, C.C.	C a	9
Groanen, S.A.R.	E s	7
Grobler (Lichtenburg), S.A.R.	F a	7
Groen Fontein, C.C.	B c	7
Groenekloof, C.C.	C f	8
Groene R. (Namaland), C.C.	B c	8
Groen R. (Victoria West), C.C.	B c	8
Groenwater, C.C.	B c	7
Groenwald, S.A.R.	C a	8
Groot Choing, Bech.	E a	7
Groot Derm, C.C.	A b	7
Groote Berg, The, C.C.	B c	8
Grootebosch, C.C.	A b	7
Groote Riet, C.C.	E b	8
Groote R. (Ceres), C.C.	E f	8
Groote R. (Ladismith), C.C.	F f	8
Groote R. (Willowmore), C.C.	C e	9
Groote River Heights, C.C.	D f	9
Groote Toorn Berg, C.C.	D d	8
Groot fontein, Bech.	A c	4
Grootfontein, C.C.	F f	8
Groot Hartz R., S.A.R.	A e	13
Groot Modder Fontein Pan, C.C.	B b	9
Groot R., C.C.	D b	7
Groot R., C.C.	A b	7
Groot Valleuvel, The, C.C.	C b	9
Gros Kraal, C.C.	C b	9
Grontville, Nat.	E d	10
Guapa Peak, C.C.	D d	15
Gustablin, Mash.	D d	15
Guay R., Mat.	C e	15
Guchas, G.S.W.A.	A b	4
Guengue, P.E.A.	C d	16
Guibis, P.E.A.	E f	15
Guigandos, G.S.W.A.	A b	4
Guingua R., C.C.	E f	8
Gulana, C.C.	C f	9
Gunyaca, P.E.A.	F f	9
Guzelschap Bank, C.C.	B b	9
Gwal R., Mat.	D e	5
Gwali, C.C.	F a	7
Gwamba, S.A.R.	F c	12
Gwena R., Mash.	E b	15
Gwobi R., Mash.	E b	15

H

HABAN, P.E.A.	F d	15
Habobes Tr., G.S.W.A.	A c	4
Haboebe Tr., G.S.W.A.	A b	4
Habooab, G.S.W.A.	A a	7
Hackney, C.C.	F c	9
Hadab, G.S.W.A.	A b	4
Hadse, S.A.R.	C e	12
Haertbertsburg, S.A.R.	E b	13
Haib, G.S.W.A.	B b	7
Haigap R., Bech.	C a	7
Haigas, G.S.W.A.	A c	4
Hakihais, G.S.W.A.	A n	7
Hakhili R., Bas.	B d	10
Halata, P.E.A.	F d	15
Halesowen, C.C.	E e	9
Halle, New, S.A.R.	D d	13
Hamburg, C.C.	G f	9
Hamles, Bech.	A c	4
Hamilton, S.A.R.	D d	13
Hamilton, Mt., Bas.	B d	10
Hamis, G.S.W.A.	B b	7
Hampden, Mt., Mash.	E b	15
Hanu, R., G.S.W.A.	B b	7
Hanah, G.S.W.A.	A c	4
Hangklip Berg, S.A.R.	D c	13
Hangklip, Cape, C.C.	C g	8
Hanjoka, C.F.S.	B b	16
Hankam, O.F.S.	B b	16
Hankey, C.C.	D f	9
Hannu Berg, G.S.W.A.	B a	7
Hanover, C.C.	D d	9
Hanover Road Sta., C.C.	D c	9
Hans Berg, The, C.C.	B a	9
Hantam Berg, The, C.C.	B c	7
Hantam East, C.C.	D d	8
Hantam, R., C.C.	B c	8
Hanyani, Mash.	E b	15
Hanyani, R., Mash.	E b	15
Hara, B.C.A.	C b	16
Haraxas Ford, S.W.A.	A c	4
Hardcastle, C.C.	D b	7
Hardeveld, The, C.C.	B d	8
Harding, Nat.	C e	10
Haris, G.S.W.A.	A b	4
Haris, S.A.R.	B a	7
Harmony, The, S.A.R.	F c	13
Harrismith, O.F.S.	C c	10
Harrison Cove, C.C.	B a	9
Harrison, Fort, C.C.	C f	10
Hartebeestfontein, S.A.R.		
Hartebeest Kraal, C.C.	B a	13
Hartebeest Mts., C.C.	D g	8
Hartebeest, R. (Great Bushman Land), C.C.	E b	8
Hartebeest, R. (Namaqualand), C.C.	C c	8
Hartebeest fontein, S.A.R.	D d	13
Hartebeesthuabm, S.A.R.	C b	4
Hartingsburg, S.A.R.	D c	13
Hartley, S.A.R.	C c	12
Hartley, Fort, Bas.	A c	10
Hartley Gold Field, Mash.	E c	15
Hartley Hill, Mash.	E c	15
Hartoch, S.A.R.	E c	12
Hart R., S.A.R.	E b	7
Haris R., C.C.	F g	8
Harts R., S.A.R.	E a	7
Hartzogs Rand, C.C.	F c	5
Hartzogs R., C.C.	F c	8
Hankein Tr., G.S.W.A.	A a	4
Haverklip, s.A.R.	D c	13
Hawston, C.C.	D g	8
Hay, C.C.	D b	7
Hebron, C.C.	F a	7
Hebron, S.A.R.	E c	13
Hebron, S.A.R.	C d	13
Hebron Road Sta., C.C.	D a	9
Heenen Woers Koppen, C.C.	E b	10
Heidelberg, C.C.	E g	8
Heidelberg, S.A.R.	D e	13
Heikab, G.S.W.A.	A b	4
Heilbron, O.F.S.	D f	13
Heilbron Road Sta., O.F.S.	C f	13
Helpmakaar, Nat.	C d	10
Helvetia, O.F.S.	F b	13
Henderson, Cape, C.C.	B g	10
Hendricks, C.C.	C g	8
Hendriksfontein, C.C.	C f	8
Hennings, O.F.S.	F a	7
Hennops R., S.A.R.	D d	13
Herakhas Ford (Orange R.), S.W.A.	C a	8
Herold, Port, Ny.	F e	3
Herrensburg, Nat.	D c	10
Hermon, Bas.	A d	10
Hermon R., C.C.	C f	8
Herschel, C.C.	G c	9

Hertzog, C.C.	F e	9
Ilex Berg, The, C.C.	B d	7
Ilex R., C.C.	D f	8
Ilex R., S.A.R.	C d	13
Hex River, town, C.C.	D e	8
Ilex River Mts.	D f	8
Hex River Sta., C.C.	D f	8
High Veld, The, S.A.R.	A f	13
Himandu, O.F.S.	B c	16
Hinga Rand, Sw.	E a	10
Hiscock, O.P.S.	G b	7
Hlabana Lake, Zul.	F e	10
Hlobane Mts., S.A.R.	D b	10
Hoaklauas, G.S.W.A.	B f	3
Hoanus, G.S.W.A.	A a	7
Honseb, G.S.W.A.	A b	4
Hodgson, S.A.R.	C c	12
Hoedspruit, S.A.R.	D c	12
Hoek Bergen, The, S.A.R.	C c	13
Hoffenland, Nat.	C c	10
Hoffman's Drift, O.F.S.	F a	7
Hofmeyer, S.A.R.	E b	13
Hogskin Vley, Bech.	B c	4
Hollbech Strand Fontein, C.C.	B c	7
Hollfontein (Potchefstroom), S.A.R.	B d	12
Hol Fontein (Rustenburg), S.A.R.	B c	12
Holgat R., C.C.	A a	8
Holle R., C.C.	F c	7
Holloway, S.A.R.	F a	7
Hololo R., Bas.	B e	10
Holspruit, O.F.S.	E b	7
Holtshausen, S.A.R.	C c	12
Hondeblats R., C.C.	D c	9
Hondeklip, C.C.	B c	8
Hondeklip Bay, C.C.	B a	8
Honde R., P.E.A.	F c	15
Honingsrestkloof, C.C.	D b	9
Honing Vley, Bech.	B c	4
Hoope Veldt, The, S.A.R.	B d	12
Hoop Point, C.C.	E b	8
Hoopstad, O.F.S.	E b	7
Hopefield, C.C.	C f	8
Hopetown, C.C.	D b	9
Horos, G.S.W.A.	A c	4
Hoshin, P.E.A.	D d	16
Hottentots Holland Mts., C.C.	C g	8
Hottentots or Koi-Koin, The, S.W.A.	A b	4
Houm Ford (Orange R.), S.W.A.	C a	8
Hout Bay, C.C.	C g	8
Houtbosch, S.A.R.	D b	12
Houtbosch Berg, S.A.R.	E c	13
Houtboschdorp, S.A.R.	E b	13
Houtkop, The, S.A.R.	B e	13
Houtkraal, C.C.	C e	9
Hout R., S.A.R.	E b	13
Houw Hoek, C.C.	C g	8
Houwwater, C.C.	B e	9
Howick, Nat.	D d	10
Howobis, G.S.W.A.	A c	4
Hubakwin, Sw.	D d	12
Huyser, S.A.R.	F a	7
Humansdorp, C.C.	D f	9
Humansdorp, dist., C.C.	D f	9
Humiasis, G.S.W.A.	A a	7
Huns, G.S.W.A.	B b	7
Hurd I., P.E.A.	D d	16
Hunh, G.S.W.A.	B a	7
Hutchinson, Nat.	C c	10
Hutub R., G.S.W.A.	A b	4
Hygap R., Bech.	B c	4

I

IBEKA, C.C.	B g	10
Ibisi R., C.C.	C e	10
Ibo, P.E.A.	G d	3
Iduchywa, C.C.	B g	10
Idumeni, Mt., Nat.	D c	10
Idutywa, C.C.	B e	7
Ifafa, Nat.	D e	10
Ifafa R., Nat.	D e	10
Ifufa, B.C.A.	A d	16
Ifumi, Nat.	D e	10
Ignambira Mt., P.E.A.	F c	15
Inasoemba R., P.E.A.	E d	15
Igutulege, P.E.A.	D c	10
Igogo, Nat.	C b	10
Igogo R., Nat.	C b	10
Igugumba, Tong	E d	12
Ihanyana, S.A.R.	D d	12
Ikhebi R., Nat.	C e	10
Ilalinbitwa R., Nat.	D d	10
Ilzer Fontein Point, C.C.	A b	4
Ikobah Tr., G.S.W.A.	A b	4
Ikogha R., C.C.	C g	10
Ikweni Lamaci, Nat.	C e	10
Ilitin, Sw.	D d	12
Ilovo R., Nat.	D e	10
Imbazani R., Nat.	D c	10
Imbrist R., P.E.A.	E c	12
Imbewula, C.C.	B f	10

Imbondune, P.E.A. ... E b 12
Imboongana, P.E.A. ... F d 15
Imbwata, C.C. ... A f 10
Imbyasuse, Mat. ... F a 10
Imihlweni, Tong. ... F n 10
Imgodini, C.C. ... G e 7
Imizizi Tr., C.C. ... C f 10
Impako, P.E.A. ... F d 16
Impanda, P.E.A. ... F d 15
Imperani, S.A.R. ... D c 12
Impiso, Sw. ... D d 12
Impogonyolo, P.E.A. ... F d 15
Impoza, P.E.A. ... D d 16
Impukani, O.F.S. ... F b 7
Impune, dist., C.F.S. ... A e 16
Imprungan, Tong. ... E d 12
Imsolan, P.E.A. ... F d 15
Imvani, C.C. ... F d 9
Imchab Berg, G.S.W.A. ... A e 4
Ina Muana, B.C.A. ... B c 10
Imanda, co., Nat. ... D d 10
Inchanga, Nat. ... D d 10
Indian Ocean, The, E.A. ... E e 16
Indina, Mat. ... E c 15
Inlowiklaue, dist., S.Z. ... E d 15
Induba, Mat. ... D c 15
Indueni, Mat. ... D d 15
Inluuduqua, Nat. ... C d 10
Inlwe, C.C. ... G d 9
Inlwe R., C.C. ... G d 9
Inembe R., Zul. ... E d 10
Infanta, Cape, C.C. ... E g 8
Ingadu Beacon, Tong. ... E a 10
Ingalele R., S.A.R. ... E a 13
Ingubini, C.C. ... C f 10
Ingogo Hill, Zul. ... D c 10
Ingome Rand, S.A.R. ... E b 10
Ingonaslang, S.A.R. ... 13A
Ingwesi R., Mash. ... E d 15
Inhacaroa, P.E.A. ... D d 16
Inhacos, B.C.A. ... D b 15
Inhantsane, P.E.A. ... F f 3
Inhauduu, P.E.A. ... D d 16
Inhanbande, S.A.R. ... G c 13
Inblazan R., Zul. ... F c 10
Inkachia, O.F.S. ... A b 16
Inkuupi R., S.A.R. ... C e 12
Inngwule, P.E.A. ... D c 16
Inowamgwane, Nat. ... C d 10
Insengalsd R., Mash. ... E b 15
Insipo, B.C.A. ... B b 10
Insuzi R., Zul. ... D c 10
Intembni, Mat. ... D c 15
Intsheeh, Tong. ... E d 12
Intombe, S.A.R. ... D d 12
Inxu R., C.C. ... B f 10
Inyack I., P.E.A. ... D c 4
Inyagu R., Mash. ... E b 15
Inyagarukulzi, R., Mash ... E b 15
Inyambe, P.E.A. ... D c 4
Inyankoyo, P.E.A. ... 3A
Inyaueni, S.A.R. ... D d 12
Inyamashenga R., Mash. ... E c 15
Inyami, P.E.A. ... D c 4
Inyangeri R., Mash. ... D d 16
Inyaugoma, I. of, P.E.A. ... D d 16
Inyangomba, P.E.A. ... D b 9
Inyasango, P.E.A. ... D c 16
Inyati, Mat. ... D c 15
Inyatsutsu, P.E.A. ... C d 16
Inyazitzi, Mash. ... F c 15
Inyawabe Mt., P.E.A. ... 3A
Inzinghazi R., Mash. ... C d 16
Ipolela, Nat. ... G b 7
Ipolela, co., Nat. ... G b 10
Ipulela R., Nat. ... C d 10
Iramba, dist., B.C.A. ... E c 16
Iramba Tribe, C.F.S. ... E e 16
Irati, Mt., P.E.A. ... D c 16
Irene Estate, S.A.R. ... D c 13
Isandhlana, S.A.R. ... D c 10
Iserwark Point, C.C. ... G e 7
Ishungwani, C.C. ... G e 7
Isibuga, P.E.A. ... E c 3
Isibugn, P.E.A. ... D b 9
Isighlul, C.C. ... F c 9
Isimleni, Mat. ... D c 15
Isipinge, Nat. ... D d 10
Island Point, C.C. ... B c 8
Iswan, R., C.C. ... D c 10
Itembeni, Nat. ... D c 10
Itepa, S.A.R. ... D b 12
Itlumel, t. C. ... E b 7
Itule, P.E.A. ... D c 4
Ityane Rock, Sw. ... G e 13
ItyenahlovuRock,S.A.R. ... D b 10
Ivuma, Mt., Nat. ... D c 10
Ixopo, co., Nat. ... D c 10
Ixopo, R., Nat. ... D c 10
Izerwark Point, C.C. ... F g 8
Izolo, C.C. ... F g 7

J

Jackals Water, C.C. ... D b 9
Jackson, Sw. ... G e 13

Jackson, Fort, C.C. ... G e 9
Jacob Reef, C.C. ... B e 8
Jacobsdal, O.F.S. ... B b 9
Jacobsdal, S.A.R. ... B c 12
Jacobskop, C.C. ... C f 9
Jacoby, S.A.R. ... C e 12
Jacomo, S.A.R. ... D b 12
Jagersfontein, O.F.S. ... E b 9
Jagtpan Rand, C.C. ... F e 8
Jakhalsfontein, S.A.R. ... B c 13
Jakhals Pts., C.C. ... A b 7
Jakhals, R., C.C. ... C e 8
James Point, P.E.A. ... E c 16
Jamestown, C.C. ... F d 9
Jamestown, S.A.R. ... G d 13
Jan Dissels, R., C.C. ... C e 8
Jangaula, C.F.S. ... B b 16
Jan Masiba, Bech. ... A c 12
Jansen, S.A.R. ... D c 12
Jansenville, C.C. ... D e 9
Jansenville, dist., C.C. ... D f 9
Janzwarts Bergen, C.C. ... E d 8
Jennette Peak, S.A.R. ... D e 12
Jellalabad, S.A.R. ... D c 12
Jellalabad, Fort, S.A.R. ... F c 13
Jeppe (Johannesburg), S.A.R. ... 13A
Jeppe (Itusidenburg), S.A.R. ... B c 12
Jeppestown, S.A.R. ... D c 13
Jericho, S.A.R. ... C d 13
Jingen, Mat. ... D c 15
Job, Tong. ... E d 12
Joho, P.E.A. ... 3A
Johannesburg, S.A.R. ... D c 13
Johannesburg, Inset map of, S.A.R. ... 13A
Johnston Falls (R. Luapula) ... D d 3
Johnston, Fort, Ny. ... F d 3
Jopo, C.C. ... G e 7
Jokesky R., S.A.R. ... C e 12
Jole R., S.Z. ... D a 15
Jonker Water, town, C.C. ... D c 7
Jordan, S.A.R. ... G a 7
Jorsberg, The, C.C. ... C b 9
J. O. Smith Bay, C.C. ... A b 7
Joubert (Barpolo), S.A.R. ... D d 12
Joubert (Middelburg), S.A.R. ... D c 12
Joubert(Pretoria),S.A.R. ... C e 12
Joubertsbnop, S.A.R. ... C c 12
Juan de Nova I., P.E.A. ... E d 16
Jumbla, Mt., C.C. ... B e 10
Jutten I., C.C. ... A d 7

K

Kaaien Veld, The, C.C. ... B b 10
Kraifontein, O.F.S. ... A b 9
Kaal Spruit, O.F.S. ... F b 9
Kaamagua, S.A.R. ... B c 12
Kaap Plateau, The, C.C. ... C a 9
Kaap Plateau, The, S.A.R. ... F d 13
Kaap R., S.A.R. ... D c 12
Kaap R., North, S.A.R. ... G d 13
Kaap R., South, S.A.R. ... F d 13
Kab, G.S.W.A. ... B a 7
Kabaer, B.C.A. ... A c 16
Kabele, C.F.S. ... A b 16
Kabinga, B.C.A. ... C c 16
Kabiskow Berg,The,C.C. ... D c 8
Kabompo, P.W.A. ... A c 16
Kabompo R., B.C.A. ... C d 3
Kaboekuk Vloer, C.C. ... E c 5
Kaboo R., C.C. ... F c 9
Kab R., G.S.W.A. ... B a 7
Kabusic R., C.C. ... A g 10
Kabusie R., C.C. ... A g 10
Kachrika, Ny. ... D c 16
Kadanga, Ny. ... D c 16
Kadzi R., S.Z. ... C b 15
Kaffir Drift, S.A.R. ... D c 13
Kaffir Kuyl Bay, C.C. ... C c 7
Kaffir Kuyl R., C.C. ... F g 8
Kaffir Pan, C.C. ... E b 9
Kaffir R., O.F.S. ... E b 9
Kaffraria, S.A. ... B f 10
Kafina, P.W.A. ... A c 16
Kafinbe, B.C.A. ... C b 16
Kafimbi, C.F.S. ... B c 16
Kafue, R., B.C.A. ... D c 3
Kafuko, C.F.S. ... B c 16
Kafumbi, B.C.A. ... B c 16
Kafundanga, P.W.A. ... A c 16
Kajanani, O.F.S. ... B b 16
Kahando, Lake, C.F.S. ... B c 16
Kahinga, B.C.A. ... B b 16
Kahinga (or Kalonko), B.C.A. ... B d 16
Kahlamba Peak, C.C. ... D e 16
Kahunda, B.C.A. ... C b 16
Kaias Mts., G.S.W.A. ... B a 7
Kainta, B.C.A. ... B c 16
Kaouko, B.C.A. ... B d 16
Kakaman, Bech. ... C b 7

Kakole, Bech. ... D a 7
Kakolole Rapids (R. Zambesi), S.A. ... C d 16
Kalabas Kraal, C.C. ... C f 8
Kalalas Pan, C.C. ... D b 9
Kalahari Desert, The, S.A. ... C f 3
Karzen, G.S.W.A. ... B a 7
Karas Berg, G.S.W.A. ... A c 4
Karawa, P.E.A. ... E c 16
Karres Kroon, C.C. ... E e 9
Karuga, R., C.C. ... F g 9
Karriga R., C.C. ... C c 9
Karena, G.E.A. ... E c 3
Kariba (George) (R. Zambesi), S.A. ... B d 16
Kariega Bosch, C.C. ... B c 7
Karima, C.F.S. ... A b 16
Karinva Fall (R. Zambesi), S.A. ... E a 15
Karkai, Bech. ... B b 4
Karoh, P.E.A. ... D c 16
Karoobasc Rapids (R. Zambesi), S.A. ... E c 3
Karola, P.E.A. ... D d 16
Karonga, Ny. ... E d 3
Karora, P.W.A. ... A a 4
Karree Bergen (Carnarvon), The, C.C. ... D d 8
Karree Bergen (Vanrhynsdorp), The, C.C. ... C d 8
Karri-Karri Salt Pan, Bech. ... C d 15
Karroo Doorn R., C.C. ... C d 8
Karroo, Moordenaars, C.C. ... F e 8
Karroo, The, C.C. ... C h 3
Karroo, The Barren, C.C. ... C c 8
Karroo, The Hokkeveld, C.C. ... D d 8
Karroo, The Great, C.C. ... D d 7
Krus R., C.C. ... D d 7
Karumbo, B.C.A. ... C c 16
Kasenga, Ny. ... G e 16
Kasanya, P.E.A. ... C d 16
Kasembe, B.C.A. ... B b 16
Kasenga, B.C.A. ... C e 16
Kasda R., B.C.A. ... B b 16
Kashoko R., C.F.S. ... A c 16
Kashull, C.C. ... G c 7
Kasiane, B.C.A. ... B b 16
Kasinde, C.F.S. ... B b 16
Kasesha, S.A.R. ... C e 13
Kassongo, dist., C.F.S. ... B b 16
Kasali, Lake, C.F.S. ... B b 16
Kassongo, P.W.A. ... A a 4
Kasteel Poort, C.C. ... D c 10
Katukosuko, Ny. ... C c 16
Kasumga, Ny. ... C c 16
Kasembe, C.C. ... D c 3
Katumamba, Ny. ... C c 16
Katanga, C.F.S. ... B d 16
Katango, Ny. ... D d 16
Katapana, C.F.S. ... B b 16
Katembe, P.E.A. ... B c 13
Katende, C.F.S. ... A b 16
Katengira, B.C.A. ... A c 16
Katere, Mash. ... E b 15
Kathlamba Mts., S.A. ... B e 10
Kathoek, C.C. ... C c 7
Katima Cataract (R. Zambesi), S.A. ... A d 16
Katima Mollo Cataract (R. Zambesi), S.A. ... B b 15
Katkop, C.C. ... E c 8
Katkop Berg, The, C.C. ... E c 8
Katlachter, S.A.R. ... D b 12
Katle, Bech. ... C c 9
Katonga, B.C.A. ... A d 16
Katonga, O.F.S. ... A d 16
Katonga, Ny. ... E c 3
Katunga, C.F.S. ... D c 16
Katundu R., S.W.A. ... B a 4
Kaviras Fall (R. Zambesi), S.A. ... B d 16
Kawala, C.F.S. ... A b 16
Kawicia, G.S.W.A. ... A c 4
Kawenga, C.F.S. ... C c 8
Kduakoro Rapids (R. Zambesi), P.E.A. ... F a 15
Keerom, C.C. ... F b 7
Keerom Berg, The, C.C. ... D f 8
Kentanicbonop, G.S.W.A. ... B g 4
Kelcuna, Bech. ... D b 7
Keighap R., C.C. ... D b 7
Kei R., Black, C.C. ... B f 10
Kei R., Great, C.C. ... B g 10
Kei R., White, C.C. ... B f 9
Keiskamahoek, C.C. ... G c 9
Keiskamma, R., C.C. ... G u 9
Kele, Mt., C.C. ... G b 7
Kembe, P.E.A. ... B c 16
Kenhardt, C.C. ... F b 8

Kenjene, C.F.S. ..... B b 16
Kentani, C.C. ..... B g 10
Kersos, G.S.W.A. ..... A e 4
Keurboom, R., C.C. ..... C f 9
Khauseh, G.S.W.A. ..... B a 7
Khama's Country, Bech. ..... B d 15
Khambes, G.S.W.A. ..... A a 7
Khams, G.S.W.A. ..... B a 7
Khamus, G.S.W.A. ..... A b 7
Khatle, Bech. ..... A e 12
Khuis, C.C. ..... B a 9
Khqu-ib R., G.S.W.A. ..... A b 4
Khoams, G.S.W.A. ..... A b 4
Khoodes, G.S.W.A. ..... A b 4
Khoros, G.S.W.A. ..... B a 7
Khosis, Bech. ..... D a 7
Khouganab, G.S.W.A. ..... B b 7
Khowos, G.S.W.A. ..... A b 4
Khuris, G.S.W.A. ..... A b 4
Kichema, P.W.A. ..... A a 4
Kikuije, C.F.S. ..... A b 16
Kikambo, Lake, C.F.S. ..... A b 16
Kibaen Mt., C.F.S. ..... B b 16
Kibuvi, C.F.S. ..... B e 16
Kifembe, C.F.S. ..... A b 16
Kifuntue, C.F.S. ..... B b 16
Kikondin, C.F.S. ..... B b 16
Kikvooneh Berg, C.C. ..... B d 9
Kimuchendo, P.W.A. ..... A e 16
Kilduncan, C.C. ..... B b 8
Kilembe, C.F.S. ..... B b 16
Kiliman, P.E.A. ..... F e 3
Kilordos, P.W.A. ..... A e 16
Kilului R., C.F.S. ..... B b 16
Kilwa I. (Lake Moero), B.C.A. ..... B b 16
Kilwa, Lake, Ny. ..... B d 16
Kimberley, C.C. ..... D a 9
Kimubere, B.C.A. ..... A e 16
Kinchewe, Bech. ..... C b 4
Kincora, O.F.S. ..... F b 9
King William's Town, C.C. ..... G e 9
Kishama, C.F.S. ..... B e 16
Kippon Point, C.C. ..... E e 7
Kirim, C.F.S. ..... B b 16
Kirk Mts., Ny. ..... C d 16
Kirui I. (Lake Bangweolo), C.A. ..... B e 16
Kirwa, C.F.S. ..... A b 16
Kisenm, B.C.A. ..... C e 16
Kisenga, P.W.A. ..... A e 16
Kisima, C.F.S. ..... B b 16
Kisima-julu Harb., P.E.A. ..... E e 16
Kisim-me, C.F.S. ..... A b 16
Kisi, R., B.C.A. ..... A e 16
Kissumbo, C.F.S. ..... A b 16
Kitangala Mts., B.C.A. ..... B e 16
Kiteraju, P.E.A. ..... E e 16
Kiteve, dist., P.E.A. ..... F e 15
Kitebe, B.C.A. ..... D a 16
Kivindu, Ny. ..... C b 16
Kivolani, P.E.A. ..... E d 16
Kiwari, C.F.S. ..... B b 16
Kiuar Fontein, C.C. ..... B d 7
Kluurstroom, C.C. ..... B f 9
Klaas Smits R., C.C. ..... F d 9
Klaver Valley, C.C. ..... B e 7
Klein Bruntjes Hoogte, C.C. ..... E f 9
Klein Choing, Bech. ..... E a 7
Klein Fristaat, dist., Sw. ..... F e 13
Klein Hartz R., S.A.R. ..... A e 13
Klein Letaba Gold Field, S.A.R. ..... F b 13
Klein Marico R., S.A.R. ..... B d 13
Klein Modder Fontein Pan, C.C. ..... B e 9
Klein Muiden, S.A.R. ..... G d 13
Klein Oliphants R., S.A.R. ..... E d 13
Klein Poort, C.C. ..... D f 9
Klein R., C.C. ..... D g 9
Klein Riet R., C.C. ..... C a 9
Klein Roggeveld, C.C. ..... E e 5
Klein Tafelberg, C.C. ..... C c 5
Klein Toorns M., C.C. ..... D d 5
Klein Vaalheuvel, C.C. ..... C b 5
Klein Winterhoek Mts., C.C. ..... D f 9
Klerksdorp, S.A.R. ..... D b 13
Klerksdorp's Drift, S.A.R. ..... B e 13
Kliphank, C.C. ..... B o 9
Klipdam, S.A.R. ..... E b 13
Klip Drift (Fraserberg), C.C. ..... E d 5
Klip Drift (Sutherland), C.C. ..... E o 5
Klipfontein, S.A.R. ..... E e 13
Klip Fontein (Calvinia), C.C. ..... C d 7
Klip Fontein (Little Namaqualand), C.C. ..... A b 7
Klip Gat, C.C. ..... B b 7
Klipheuvel, C.C. ..... C f 5
Klip Kuil, Bech. ..... B b 7
Klip Kuil, S.A.R. ..... B e 12

Klippen Point, C.C. ..... D g 9
Klipphant, C.C. ..... D e 5
Klip R., Nat. ..... C e 10
Klip R., S.A.R. ..... G a 7
Klip R. (Heilbron), O.F.S. ..... D f 13
Klip R. (Vrede), O.F.S. ..... E f 13
Klip River, co., Nat. ..... C e 10
Klip River, C.C. ..... B c 7
Klip Stapel, S.A.R. ..... D d 12
Knaas, Bech. ..... B c 4
Knapzak R., O.F.S. ..... B e 9
Knolnoses, The, S.A.R. ..... F b 13
Knysna, dist., C.C. ..... C f 9
Knysna Harb., C.C. ..... B g 9
Koba, B.C.A. ..... B d 16
Koiamshodi Kopjes, The, S.A.R. ..... G c 13
Kobbies Berg, The, C.C. ..... B c 7
Kobe, Bech. ..... B d 15
Kobis, Bech. ..... B b 4
Kobis, G.S.W.A. ..... A a 4
Koe Berg, The, C.C. ..... C b 8
Koednes Berg, The, C.C. ..... C d 7
Koegas, C.C. ..... B b 9
Koesterfontein (Gold Field), S.A.R. ..... B d 13
Koffiefontein, O.F.S. ..... B e 9
Koffiekuil, O.F.S. ..... D b 9
Kogani, Zul. ..... D d 12
Kogel Bay, C.C. ..... C g 8
Kogelbwen, C.C. ..... D b 7
Kogel Berg, The, C.C. ..... C g 8
Kogna R., C.C. ..... C d 7
Koi-Koin, The, S.W.A. ..... A a 4
Koina, G.S.W.A. ..... B a 7
Kokalu Pits, Bech. ..... B b 4
Kokenaop, C.C. ..... B c 7
Kokstad, C.C. ..... C e 10
Kokumbene, R., S.A.R. ..... G a 13
Kokuming, Bech. ..... E a 7
Kokwe, Bech. ..... E a 7
Kolberg, C.C. ..... B b 4
Kolobeng, Bech. ..... A c 12
Komaguas, C.C. ..... B b 8
Komati Gold Field, Sw. ..... G e 13
Komati Middel, S.A.R. ..... D e 13
Komati Poort, S.E.A. ..... D d 13
Komati R., S.E.A. ..... F d 13
Kombe, Zul. ..... D e 10
Komhuus, Mash. ..... E e 15
Komhuis, Mt., C.C. ..... C e 10
Komghu, C.C. ..... A g 10
Kompie R., S.A.R. ..... E c 13
Koms Berg, The, C.C. ..... E e 8
Kondivizoo Rapid (R. Zambesi), P.E.A. ..... F a 15
Kome Mts., C.F.S. ..... B e 16
Kongaberg, Nat. ..... C b 10
Koni Mt., C.F.S. ..... B c 16
Koning, Bech. ..... D a 7
Koodoes R., S.A.R. ..... F b 13
Koodloo, C.C. ..... D a 9
Koodoo Rand, The, C.C. ..... B e 13
Kooigoe I Flats, C.C. ..... D e 8
Kookfontein, C.C. ..... B b 8
Koonap R., C.C. ..... F d 9
Koopman, C.C. ..... E b 9
Koopmansfontein, C.C. ..... D a 9
Kopa, B.C.A. ..... C e 16
Kopje Alleen, The, O.F.S. ..... F a 9
Kopje Enkel, The, S.A.R. ..... A d 13
Kopjes Dam, C.C. ..... B e 9
Kopong, Bech. ..... B e 12
Koran, C.C. ..... E c 9
Korana, The, C.C. ..... F a 9
Koran Ford (Orange R.) ..... B b 7
Korannu Land, Bech. ..... B e 9
Kornet Sprnt, Res. ..... A e 9
Kornet Spruit, dist., Bas. ..... A e 10
Kurt Erasmus, S.A.R. ..... D d 13
Kosi, Tong. ..... E a 10
Kosi Bay, Tong. ..... E d 12
Kosi, Lake, Tong. ..... E d 12
Kotakota, Ny. ..... C c 16
Kouga Berg, The, C.C. ..... C d 9
Kouga R., C.C. ..... B f 9
Kouga Rand, The, C.C. ..... C f 9
Kouws Berg, The, C.C. ..... B f 9
Kowamba, Lake, C.F.S. ..... B b 16
Kowedi, C.F.S. ..... B b 16
Kowirwi Mt., Ny. ..... C c 16
Kowisin Tribe, G.S.W.A. ..... A b 4
Kraaibosh, C.C. ..... D b 9
Kraal R., C.C. ..... G c 9
Kraal R., C.C. ..... F c 9
Kraus Berg, S.A.R. ..... C c 13
Kransfontein, O.F.S. ..... B c 10
Krantskop, The, C.C. ..... B c 10
Krans Kop, The, O.F.S. ..... C b 10
Krans Kop, The, Nat. ..... D e 10
Kras-kuil, C.C. ..... D b 9
Kromellehoog R., O.F.S. ..... C e 13
Krom R., C.C. ..... D g 9
Kroon R., S.A.R. ..... D d 13
Kroonstad, O.F.S. ..... D g 13
Kroomstad, S.A.R. ..... C d 13
Krugen, S.A.R. ..... C d 13
Kruger Kraal, C.C. ..... E c 7

Krugersdorp, S.A.R. ..... C e 13
Kruger's Post, S.A.R. ..... F c 13
Kruis Fontein, C.C. ..... D f 9
Kruis R., C.C. ..... C d 9
Kruis River, town, C.C. ..... A f 9
Kudubeni, C.C. ..... A f 10
Kulbosis Berg, C.C. ..... B a 8
Kuikuins, G.S.W.A. ..... A a 4
Kuils Rand, C.C. ..... B e 9
Kuis, Bech. ..... B c 4
Kulais, Bech. ..... D a 7
Kumadan Lake, Bech. ..... B d 15
Kumana Location, S.A.R. ..... A c 13
Kump R., C.C. ..... F c 9
Kunda, B.C.A. ..... E d 3
Kunene, R., W.A. ..... A e 3
Kungwe, B.C.A. ..... C e 16
Kunuwa, Bech. ..... B b 12
Kunwana, S.A.R. ..... A d 12
Kunyara, Bech. ..... B d 15
Kura Hills, P.E.A. 3A
Kurrabella, Bech. ..... B b 12
Kuruman, Bech. ..... D a 7
Kuruman R., Bech. ..... D a 7
Kushito, B.C.A. ..... B c 16
Kutling, Bas. ..... F c 7
Kutling, dist., Bas. ..... A e 10
Kutling R., Bas. ..... B e 10
Kws Kwa R., P.E.A. ..... F c 3
Kwamaguaza, Zul. ..... E c 10
Kweio R., P.W.A. ..... B a 4
Kwelegha Point, C.C. ..... B g 10
Kwelegha R., C.C. ..... C d 7
Kwilo R., P.W.A. ..... A a 4
Kybaka's Pan, Bech. ..... C c 15
Ky Gariep R., S.A. ..... G a 7

L

LADISMITH, C.C. ..... F f 5
Ladybrand, O.F.S. ..... G b 9
Lady Frere, C.C. ..... G d 9
Lady Grey (Aliwal North), C.C. ..... G c 9
Lady Grey (Parel), C.C. ..... C f 9
Lady Grey (Robertson), C.C. ..... D f 8
Lady Koch, The, C.C. ..... B c 10
Ladysmith, Nat. ..... C c 10
Lahombo, Mash. ..... D d 15
Laingsburg, C.C. ..... F f 5
Laing's Nek, Nat. ..... C b 10
Laken Valley, C.C. ..... C d 9
Lakersing, C.C. ..... A b 7
Lake Matata, C.C. ..... D e 16
Lambert Bay, C.C. ..... C e 8
Langdaan, C.C. ..... C f 5
Lange Berg, The, C.C. ..... B c 7
Lange Bergen, The, Bech. ..... B b 9
Lange Bergen (Namaqualand), The, C.C. ..... C c 8
Lange Berge (Swellendam & Riversdale), C.C., The, ... E f 5
Lange Kuit, C.C. ..... E d 8
Langkloof, C.C. ..... B a 9
Langkloof Mts., C.C. ..... C f 9
Lang Kloof R., C.C. ..... A e 9
Lang's Nek, Nat. ..... G a 7
Langspruit, O.F.S. ..... E e 13
Larcon, S.A.R. ..... E c 13
Lat Lake Kraal, Bech. ..... B b 4
Leadsman Shoal, Tong. ..... F b 10
Le Blan Falls (Maletsunyana R.), Bas. ..... B d 10
Lebanami Pool, Bech. ..... D a 7
Leebanas, C.C. ..... C d 4
Ledinguani, Bas. ..... B d 10
Leeuw Spruit, S.A.R. ..... E a 7
Leeuw Kop, The, S.A.R. ..... E a 7
Leeuwen Drift, C.C. ..... C c 7
Leeuwen Kuit, C.C. ..... B c 7
Leeuwfontein, C.C. ..... B d 9
Leeuw Klip, The, C.C. ..... C b 5
Leeuw R., C.C. ..... B c 9
Leeuw R. (Ladybrand), O.F.S. ..... F b 9
Leeuw R. (Vrede), O.F.S. ..... F b 10
Legobatse, Bech. ..... D a 7
Lekonic R., B.C.A. ..... B d 16
Lekuni, Bas. ..... B d 10
Lelintitung, Bech. ..... D b 4
Lemondo, S.A.R. ..... D b 13
Lepahole R., S.A.R. ..... D e 13
Lepata, S.A.R. ..... G b 13
Leporo, S.A.R. ..... B c 12
Leritu, Bas. ..... B c 10
Lerin, dist., Bas. ..... A d 10
Lerothondi, Bas. ..... A d 10
Lesatsitche, Bech. ..... A d 15
Leshoboro, Bas. ..... A d 10
Lesimatche's, Bech. ..... A d 15
Le Souvenir, O.F.S. ..... G b 9
Losseytown, C.C. ..... F d 9

Letaba, Bas. ..... B d 10
Letaba R., Great, S.A.R. ..... F b 13
Letaba R., Klein, S.A.R. ..... E b 13
Lerbabat R., Midden, S.A.R. ..... F b 13
Letebo, Mat. ..... C c 15
Letjesbosch, C.C. ..... E e 9
Letloche, Bech. ..... C e 15
Letsea R., Bas. ..... F b 7
Letsen's, Bas. ..... B d 10
Letsiielo R., S.A.R. ..... D b 12
Leuwdeorn, S.A.R. ..... F a 7
Leuwen Drift, C.C. ..... E d 8
Leven Drift, C.C. ..... F g 8
Leven Point, Tong. ..... F b 10
Levubo R., S.A.R. ..... F a 13
Leydsdorp, S.A.R. ..... F c 13
Lialui, B.C.A. ..... C e 3
Liambai R., B.C.A. ..... C d 3
Liana R., P.W.A. ..... B a 4
Libako I., B.C.A. ..... A d 16
Liba R., B.C.A. ..... C d 3
Libanga R. P.W.A. ..... A c 16
Libata, C.F.S. ..... B c 16
Libatas, B.C.A. ..... B c 16
Lichtenburg, S.A.R. ..... B e 13
Lienogo, P.E.A. ..... D d 16
Lidgettown, Nat. ..... D d 10
Liebensberg's Vlei R., O.F.S. ..... B b 10
Lieuw Berg, The, C.C. ..... C b 9
Lifungo I. (Lake Bangweolo), C.A. ..... B c 16
Ligonis R., C.F.S. ..... D d 16
Li'anblong, C.C. ..... E b 7
Likoma, P.E.A. ..... E d 3
Likondo, C.F.S. ..... A b 16
Lakoto, P.E.A. ..... E c 12
Lilyfontein, C.C. ..... C c 8
Limpopo R., S.A.R. ..... C b 12
Linuvuba R., S.A.R. ..... D b 4
Lindley, O.F.S. ..... E e 13
Lingalo, Bech. ..... A c 16
Linokans, S.A.R. ..... B c 12
Lintjes R., Kwari, C.C. ..... B c 8
Linyanti, G.S.W.A. ..... A c 15
Lion Pan, G.S.W.A. ..... B a 4
Lion R., Great, C.C. ..... B f 9
Lions River, co., Nat. ..... C d 10
Lishebe, P.E.A. ..... D c 16
Lister, Fort, Ny. ..... D d 16
Lisunga, B.C.A. ..... C c 16
Lisane R., S.A.R. ..... G a 13
Litsela, B.C.A. ..... B c 16
Liteyana, Bech. ..... A c 12
Lidatlong, C.C. ..... D a 9
Litofe, B.C.A. ..... A d 16
Little Brak R., C.C. ..... F d 8
Little Bushman Land, C.C. ..... B c 8
Little Caledon R., O.F.S. ..... B c 10
Little Fish R., C.C. ..... E e 9
Little Kenns Berg, C.C. ..... C c 5
Little Lebuta R., S.A.R. ..... D b 12
Little Namaqualand, C.C. ..... A b 7
Little Pella, C.C. ..... C b 8
Little Riet R., C.C. ..... E c 9
Little Tugela R., Nat. ..... C c 10
Little Zwarte Bergen, The, C.C. ..... B f 9
Livingstonia, Ny. ..... C c 16
Liwele, C.F.S. ..... B b 16
Lhuwarme, S.A.R. ..... E f 13
Loanda, B., B.C.A. ..... B d 16
Loane, P.E.A. ..... D d 16
Loangiuga. R., B.C.A. ..... C c 16
Loangwa R., B.C.A. ..... C c 16
Loangwa R., Rapids of the, B.C.A. ..... C c 16
Loanja R., B.C.A. ..... B b 15
Loano, P.E.A. ..... D d 16
Lobkrau, Bech. ..... B b 12
Lobatani, S.A.R. ..... B b 12
Lobengula, dist., B.C.A. ..... C c 16
Lobe R., B.C.A. ..... C c 16
Lobetlhal, B.C.A. ..... D c 12
Loblani, dist., B.C.A. ..... B c 12
Lobombo Mts., P.E.A. ..... E d 12
Loenge R., B.C.A. ..... B c 16
Lofu R., B.C.A. ..... C b 16
Loggane, Bech. ..... A c 12
Logge R., B.C.A. ..... C b 15
Loi, R., B.C.A. ..... A d 16
Lokera, B.C.A. ..... B r 16
Lokinga Mts., C.A. ..... B c 16
Lomati R., P.E.A. ..... E c 16
Lo Magondi's, Mash. ..... B d 16
Lo Magondi's Gold Field, Mash. ..... E b 15
Lomami R., C.F.S. ..... A b 16
Lomani R., S.A.R. ..... C c 16
Lombard R., B.C.A. ..... G a 7
Lombe R., B.C.A. ..... A a 16
Lumbu, P.E.A. ..... E c 16
Lomwe Tr., P.E.A. ..... D c 16
Longamo, B.C.A. ..... C c 16
Longamo, N.Z. ..... B c 16

Name	Ref	Pg
Longhope, C.C.	E e	9
Long Kloof, The, C.C.	D e	8
Longhuis, C.C.	D a	1
Longo R., S.W.A.	A d	16
Looking, Bech.	D b	7
Lopepe, Bech.	B b	12
Lorenzo Marques, P.E.A.	E g	3
Losanga, R., C.P.S.	B b	16
Losberg, S.A.R.	F a	7
Lositi R., B.C.A.	B d	10
Loskop, C.C.	B d	7
Los Kop, The, C.C.	E b	7
Loskop, The, S.A.R.	C e	13
Lospers Plaats, C.C.	B e	8
Lotlil, Sw.	D d	12
Lotlakana, Bech.	A d	12
Lotlokane, Bech.	B d	15
Lotsani R., Bech.	C e	15
Lot's Pillar, O.F.S.	A e	16
Louisfontein, C.C.	B e	8
Lovedale, C.C.	F e	9
Lovel, R., C.P.S.	D b	16
Lowe, C.F.S.	A b	16
Lower Drift, Nat.	E d	10
Lower Tugela, co., Nat.	E d	10
Lower Umfuli Goldfield, Mash.	D b	15
Lower Umvolosi, dist., Zul.	F c	10
Luabo, West, P.E.A.	F a	3
Luadaba, R., C.F.S.	F e	16
Luanje, P.W.A.	A a	4
Luepula R., C.P.S.	B c	16
Lubiansi R., C.F.S.	A b	16
Lubli Pits, Bech.	B b	4
Luburi R., C.F.S.	B b	16
Luchulingo Valley, P.E.A.	D e	16
Ludlow, C.C.	E d	9
Ludwigslust, S.A.R.	D e	12
Lufira, R., C.F.S.	D e	8
Lufubu, R., C.F.S.	B e	16
Luia, R., Mash.	E b	15
Luia, R., P.E.A.	C c	16
Lui-banda Tribe, P.W.A.	A e	16
Luiji, R., C.F.S.	A b	16
Luisi R., Bech.	C e	15
Luis, R., Mash.	C d	16
Luitakwe, Bech.	B e	12
Lujenda, R., P.E.A.	F d	3
Lukasap, R., C.F.S.	A b	16
Lukenkwe, R., P.E.A.	D d	16
Lukuga R., C.F.S.	D e	3
Lukuga, R., P.E.A.	D d	16
Lukungu, R., C.F.S.	A b	16
Lukwasui, R., B.C.A.	C e	16
Luli, P.E.A.	F d	3
Luli, R., P.E.A.	D e	16
Lulua, R., C.F.S.	A b	16
Lulu Mts., S.A.R.	E c	13
Lumbo, P.E.A.	F c	16
Lumessi, P.E.A.	C c	16
Luneaja, C.F.S.	A b	16
Lunur, Mash.	E d	15
Lundi, R., Mash.	E d	15
Luneberg, S.A.R.	F f	13
Lunga Mandi, C.F.S.	B b	16
Lunga, R. (R. Kabompo), B.C.A.	A e	16
Lunga, R. (Kitangula Mts.), B.C.A.	B e	16
Lungo Mashimba, C.F.S.	B b	16
Langunje R., B.C.A.	B c	16
Lupampa Mt., P.E.A.	D c	16
Luqanda, C.F.S.	A b	16
Lupata Gorge, P.E.A.	C d	16
Lurilopepe, Bech.	B e	15
Luria Bay, P.E.A.	E c	16
Lurio, R., P.E.A.	C d	16
Lusheri, C.F.S.	A b	16
Lushunu, S.Z.	B d	10
Lusiti R., P.E.A.	F c	15
Luttig, C.C.	B e	9
Luvigo R., C.F.S.	B b	16
Luweenbi R., C.F.S.	A b	16
Luwimbi, C.F.S.	E c	16
Luzizi, C.C.	B g	10
Lydenburg, S.A.R.	F d	13
Lydenburg, dist., S.A.R.	F c	13

M

Name	Ref	Pg
Maamya, Ny.	D d	16
Mabalela, C.F.S.	A c	16
Mabakatwani, Mat.	D d	15
Mabal, S.A.R.	B d	12
Mabams, S.A.R.	D e	13
Mabare Mt., S.Z.	D b	15
Mabare R., S.A.R.	D d	16
Mabeki R., C.C.	G e	7
Mabembane, S.A.R.	D e	12
Mable's Kraal, S.A.R.	B d	13
Mabo, P.E.A.	F c	15
Mabola, O.F.S.	G b	9
Maboongatajaba, S.A.R.	F b	13
Mabsa, Bech.	C f	3
Macasule, S.A.R.	D e	12
MacDougall Harbour, C.C.	A b	8
Macbaleng Flats, Bech.	B e	15
Machabe R., Bech.	A e	15
Machacha, Mt., Bas.	A d	10
Machakod, S.A.R.	D e	12
Machaquene, P.E.A.	G a	13
Machay, S.A.R.	G d	13
Macibi, Mt., C.C.	C f	10
Maclase, Bech.	A e	12
Maclean, C.C.	A g	10
Macicar, C.C.	B f	10
Macloutsie, Mat.	D d	15
Marloutsie R., Mat.	D d	15
Macoocoue, P.E.A.	H d	13
Madagascar Reef, C.C.	G f	9
Madan, P.E.A.	D e	4
Madebing, Bech.	D n	7
Madembe, B.C.A.	C b	15
Madenasama, Bech.	B e	15
Madicune, P.E.A.	F e	15
Madodo, P.E.A.	D e	16
Madrara, P.E.A.	C e	16
Madundsi, P.E.A.	F d	15
Madunjo R., P.E.A.	H b	13
Mafamede I., P.E.A.	E d	16
Mafeking, Bech.	D g	3
Mafeking R., Bech.	B e	4
Mafeteng, Bas.	A d	10
Mafogwan, P.E.A.	C c	16
Mafokone, S.A.R.	D e	16
Mafungulani Hills, S.Z.	D b	15
Mafussi, P.E.A.	F d	15
Mafutane, Sw.	D d	12
Magalaqueen R., S.A.R.	D a	13
Magalies Berge, S.A.R.	C d	13
Magalies, R., S.A.R.	C d	13
Magunges Kraal, S.A.R.	G e	13
Magato, S.A.R.	E a	13
Maghamba, S.Z.	C d	16
Maghunda, Mash.	D b	15
Magiga, Sw.	E a	10
Magne Pool, Bech.	C d	5
Magogomela, P.E.A.	F d	15
Magesi, P.E.A.	F e	8
Maguuta, P.E.A.	F c	15
Maguire Fort, Ny.	D e	16
Maguival, P.E.A.	D d	16
Magumba, G.S.W.A.	B a	4
Magumba, S.Z.	F b	16
Mahalopsi R., Bech.	B b	12
Mahntu I., P.E.A.	E c	16
Mahilaan, P.E.A.	D e	4
Mahila's Kop, S.A.R.	E b	13
Mahti Chwane, Bech.	C e	15
Maholu, Mat.	D e	15
Mahonta I., S.Z.	A d	16
Mahenti, P.E.A.	F e	15
Mahue Ma Sinique Mts., Mash.	F c	15
Mahunga, P.E.A.	F d	15
Mabun Mahbi, Bech.	B e	15
Maihonda, Bech.	C c	15
Maila, Bech.	B a	4
Maili, G.S.W.A.	B a	4
Main Dr., Nat.	C e	10
Maintirano, Mad.	G o	3
Maitland, C.C.	C f	8
Majane, Bech.	B e	15
Majeeba R., B.C.A.	B b	15
Majuba Hill, Nat.	G a	7
Makabeng, S.A.R.	C b	12
Makalaka Tr., B.C.A.	C b	15
Makambuvi, P.E.A.	D c	16
Makanda, P.E.A.	C a	15
Makandoa, P.E.A.	D e	16
Mahanga Tr., P.E.A.	D d	16
Makanjera, P.E.A.	D d	16
Makanjila, Ny.	F d	3
Makao, S.A.R.	C e	13
Makapofo Vlei, Bech.	A d	15
Makararia, P.E.A.	A e	16
Makari Kari (Salt Pan), Bech.	C d	15
Makeng, Bech.	A d	15
Makata, Bech.	A d	15
Maketta, P.E.A.	D d	16
Makhaleng Spruit, Bas.	A d	10
Makhobe, Tati	C d	15
Makhosi, S.A.R.	B c	12
Makhika, Bech.	B e	15
Makonvela, S.A.R.	F a	13
Makoc, R.C.A.	C b	15
Makoe, R., B.C.A.	C b	15
Makololo Tr., Ny.	C c	16
Makonde, B.C.A.	C b	15
Makonya Mts., Sw.	C d	16
Makopan, S.A.R.	C e	12
Makopi, R., P.E.A.	D d	16
Makori, Mash.	E e	15
Makorokoi Tr., S.Z.	C b	15
Makosini Raml, The, S.W.	E b	10
Makuatoe Mt., Bech.	C d	15
Makua Tr, P.E.A.	F e	16
Makulan, P.E.A.	F d	15
Makuleke, S.A.R.	G a	13
Mukulusina, S.A.R.	D d	12
Makwara Tr., Mat.	D b	4
Makwap'de, S.A.R.	G b	12
Makwarga, P.E.A.	F d	15
Makwassie Berge, S.A.R.	B f	13
Makwassie Spr., S.A.R.	B f	13
Malngas, C.C.	E g	8
Malan, S.A.R.	D d	12
Malang, S.A.R.	C e	12
Malans, S.A.R.	D e	12
Malema, R., P.E.A.	D e	16
Malemba, B.C.A.	C c	16
Malemboka, C.F.S.	B c	16
Malesso Mmuo, P.E.A.	D d	16
Maletsunyane R., Bas.	B d	10
Malibetse, S.A.R.	D e	12
Malinda, Sw.	D d	12
Malip, S.A.R.	E c	13
Malisa, Mnt.	C c	15
Malitzi Tr., S.A.R.	K b	13
Malmanie Gold Field, S.A.R.	A d	13
Malmani R., S.A.R.	C e	13
Malmesbury, C.C.	C f	8
Malobocha, S.A.R.	C b	12
Malo I. (R. Shiré), B.C.A.	D d	16
Maluk, S.A.R.	C e	12
Malops, P.E.A.	D d	16
Malopo R., Bech.	E a	8
Malton, Nat.	D d	10
Maluti Mts., Bas.	A d	10
Mamachah, S.A.R.	C e	12
Mamakhnisie, Bech.	B b	4
Mamatson's, Mash.	E b	15
Mamba, B.C.A.	D b	15
Mambe, S.Z.	D a	15
Mambedilo, S.A.R.	G d	13
Mamii, S.A.R.	D d	12
Mambirana Falls (R. Loapula), C.F.S.	B c	16
Mo-mburuna, B.C.A.	C d	16
Mambwe, dist., B.C.A.	C b	16
Mamele, G.S.W.A.	A b	15
Mamheny, P.E.A.	E d	12
Mamilanga, B.C.A.	B b	15
Mamochi Kraal, S.A.R.	C f	8
Mampoa, S.A.R.	E a	7
Mamuzulu, Sw.	D d	12
Manab, Tong.	E d	12
Manaba, Tong.	F b	16
Mana, R., C.F.S.	B b	16
Mananga Mt., Sw.	G e	13
Manansa Tr., Bech.	B e	15
Man-bunda, R., B.C.A.	A e	16
Manenga Tr., Ny.	C e	16
Manda, Sy.	C b	16
Mandabare, B.C.A.	B d	15
Mandala, Ny.	C d	16
Mamlanga Tr., P.E.A.	F d	15
Mandingo, P.E.A.	F b	16
Mandoya Tr., P.E.A.	F d	15
Manevring, Bech.	D a	7
Manenko's, P.W.A.	A e	16
Manganja Tr., Ny.	E d	3
Manganzara, C.C.	F o	9
Mangedge, P.E.A.	D e	16
Mangoamai, Mash.	E e	15
Mangneto, S.A.R.	F a	13
Mangundwane, P.E.A.	D d	16
Mangola, P.W.A.	A e	16
Mangwa Sensa, C.F.S.	B b	16
Mangwe, Mat.	C d	15
Mangwe R., Tati	C d	15
Manhissa R., P.E.A.	C a	15
Manica, dist., B.C.A.	C a	15
Manica, Mash.	E e	15
Manica Gold Field, E.A.	F c	15
Manica Tr., B.C.A.	B c	15
Manisane, S.A.R.	D e	16
Mankambira, Ny.	C e	16
Mankelekete, Tr., S.A.R.	G e	13
Mankoe Tr., B.C.A.	A e	16
Mankopan, S.A.R.	C b	12
Mankutane, Bech.	A e	12
Manjobo, P.E.A.	D d	16
Mano, Ny.	C c	16
Mano, dist., P.E.A.	C c	16
Manowa's, P.E.A.	D e	16
Mantanyama, B.C.A.	C b	16
Mampa, Bech.	A c	15
Manuele R., S.Z.	D b	15
Manyame, P.E.A.	C d	16
Manyane, P.E.A.	C d	16
Manyamb'we, S.A.R.	G a	13
Manueli, P.E.A.	D e	16
Mapashleka, S.A.R.	E c	13
Mapeka, S.A.R.	D b	13
Maping, Bech.	D a	7
Mapoch's, S.A.R.	E e	13
Maponda's, Mash.	E b	15
Mapota, P.E.A.	D c	16
Mapotshan, Tong.	E d	12
Mapstu, Tong.	E d	12
Mapui R., Mat.	E b	15
Maputo, B.C.A.	B c	16
Mapuza R., Sw.	D c	16
Marabastad, S.A.R.	E b	13
Maraben, S.A.R.	C b	12
Marafato, P.E.A.	E c	16
Maraus, S.A.R.	D e	12
Marais, P.E.A.	B e	13
Maraisburg, C.C.	E d	9
Marakalatu Mt., Bech.	B b	12
Maranda, B.C.A.	C e	16
Maranquan, S.A.R.	C e	13
Marwi Tr., P.E.A.	C c	16
Marburg, Nat	D e	10
Marcus Bay, C.C.	C e	7
Marelaneng, Bech.	D e	7
Maremane, Bech.	D b	7
Marembe, P.E.A.	D d	16
Marenga, Ny.	C e	16
Mareybeng, C.C.	D e	4
Mareybeng, C.C.	E a	7
Maribogo, Bech.	E a	7
Marico, dist., S.A.R.	A d	13
Marico Drift, S.A.	B e	12
Marico, R., S.A.R.	B e	13
Marico R., Klein, S.A.R.	B d	13
Mariep, S.A.R.	E a	13
Marikele Mts., S.A.R.	C e	12
Marinba, B.C.A.	B d	16
Marimba Tr., B.C.A.	B d	16
Maritsana R., Bech.	A e	12
Marlow, C.C.	E e	9
Marnewyk, O.F.S.	F a	7
Maroblang, Bech.	E a	12
Marshall, S.A.R.	B e	12a
Marshall, Fort, S.A.R.	D e	10
Martha Point, C.C.	E e	7
Martin Rock, C.C.	B e	8
Marukutu, B.C.A.	C b	16
Marule, S.A.R.	F e	13
Marutse and Malombe, Kingdom of, B.C.A.	A d	16
Masabango, Ny.	D e	16
Masamba, Mash.	E e	15
Masanji, P.E.A.	C c	16
Masarwa Tr., Bech.	B b	4
Masasina Bay, P.E.A.	E c	16
Masasai, C.F.S.	E b	16
Masceba, P.E.A.	C d	16
Maseke, S.A.R.	C b	13
Maseppa, Bech.	B e	4
Maseru, Bas.	A d	10
Masesa, C.F.S.	B b	16
Mashabba, S.Z.	C e	13
Mashango, B.C.A.	B b	16
Mashato, B.C.A.	B e	16
Mashegashe R., Mash.	E e	15
Mashoke R., Mash.	E e	15
Mashookane, S.A.R.	C b	12
Mashinalala Mts., S.A.R.	F e	13
Mashinga, dist., P.E.A.	F b	15
Mashiromba, P.E.A.	E b	15
Mashone, P.E.A.	C e	16
Mashonaland, S.A.	D b	15
Mashna, Bech.	C b	4
Maslue, Bech.	E b	12
Mashukulumbwe Tr., B.C.A.	B d	16
Mashubwa Bay, P.E.A.	D d	16
Masiringi, Mat.	D d	15
Masitisi, Bas.	A e	10
Masokalm, G.S.W.A.	B b	15
Massara Poort, S.A.R.	E b	13
Massape, P.E.A.	F e	15
Massomo, S.A.R.	D e	12
Masopha, Bas.	A d	10
Masqua, Bech.	A c	15
Ma Tabbm, S.Z.	B b	4
Matabelaland, S.A.	D e	15
Matatlin, S.A.R.	C d	13
Matalile, Ny.	C d	16
Matakana, P.E.A.	E a	15
Matakanya's, P.E.A.	C d	16
Matako R., G.S.W.A.	A a	4
Matala Poort, S.A.R.	D e	12
Matalin Poort, P.E.A.	D d	13
Matambwi Tr., P.E.A.	D e	16
Matamini, P.E.A.	D e	16
Matanda, B.C.A.	B e	16
Matangeni, S.A.R.	E e	16
Matasie, C.C.	B e	10
Matchidindo, B.C.A.	C b	16
Matchutseneng, Bch.	G b	7
Matela, Bas.	G b	7
Matemene, P.E.A.	D d	16
Matene I., P.E.A.	C e	16
Matewane Mts., C.C.	E f	9
Mathaba, Bas.	B d	10
Mathulane, Bech.	B d	15
Matibesi, P.E.A.	E e	16
Matificr, P.E.A.	F d	15
Matibi's, S.A.R.	C b	12
Matietsie R., Bech.	C e	15
Matikwiti, P.E.A.	D e	16
Matimgi, Tong.	E d	12
Matiopa, Mat.	D b	4
Matita, Ny.	C d	16

Matiti, P.E.A. ... Fd 15
Matjes Fontein (Graaf Reinet), C.C. ... Ec 7
Matjes Fontein (Hopetown), C.C. ... Dc 7
Matjesfontein (Worcester), C.C. ... Ef 4
Matlabas R., S.A.R. ... Bc 12
Matlaule R., S.A.R. ... Gc 13
Matlopine, Bech. ... Ad 12
Matlwaring R., Bech. ... Da 7
Matok, S.A.R. ... Eb 13
Matongo, B.C.A. ... Bc 16
Matope, Ny. ... Ec 3
Matopi's, Mash. ... Eb 15
Matoppo Mts., Mat. ... Dd 15
Matroos Bay, C.C. ... Ad 4
Matsap, C.C. ... Bc
Matsieng, Bas. ... Ad 10
Matsoku R., Bas. ... Bd 10
Matsopong, S.A.R. ... Cb 12
Matawanakaba, Bas. ... Be 10
Matua, B.C.A. ... Bd 16
Matumbo, P.W.A. ... Aa 4
Matuna, B.C.A. ... Ce 16
Matyatyo, S.A.R. ... Db 12
Maua Tr., P.E.A. ... Dc 16
Maubaan, S.A.R. ... Cc 12
Maueli Berg, S.A.R. ... Fd 13
Maundo Mt., B.C.A. ... Bd 16
Mavin Tr., P.E.A. ... Dc 16
Mavtti Tr., Ny. ... Cb 16
Mavongo, P.E.A. ... Ec 16
Mavus R., P.E.A. ... Cd 16
Maxongos Hoek, C.C. ... Af 10
Mayapi Bay, P.E.A. ... Gd 3
Mayorokokoro, C.C. ... Bc 4
Maytengne R., Bech. ... Cd 16
Mazamamba, B.C.A. ... Cc 16
Mazaza, Ny. ... Ce 16
Mazazima Bay, P.E.A. ... Gd 3
Mazel Fontein, C.C. ... Bc 4
Mazeppa Point, C.C. ... Bg 10
Mazeppa R., C.C. ... Bg 10
Mazimlnnguga's, Mash. ... Db 15
Mazingani R., Mash. ... Ed 15
Mazoe Gold Field, Mash. ... Eb 15
Mazoe R., Mash. ... Eb 15
Mazungwa, Mt., Mash. ... Fb 15
Mbadu, P.E.A. ... Dd 16
Mbai R., B.C.A. ... Cc 16
Mbalipi, P.E.A. ... Dc 16
Mbamea, P.E.A. ... Ce 16
Mbauza, P.E.A. ... Dc 16
Mbekeleweni, Sw. ... Dd 12
Mbewe, Ny. ... Cd 16
Mbonda Mts., P.W.A. ... Ac 16
Mbopo, B.C.A. ... Bb 16
Mbopo, B.C.A. ... Bd 16
Mbota, P.E.A. ... Ce 16
Mbuii, C.C. ... Fa 7
McArthur, S.A.R. ... Dd 13
McDermot, S.A.R. ... Fa 7
McFarlin, C.C. ... Da 9
Mchilimba, P.E.A. ... Dc 16
Mcwasa R., C.C. ... Cg 10
Mdjewah, P.E.A. ... Ed 16
Medingen, S.A.R. ... Fb 13
Medingwedingwe R., P.E.A. ... 3a
Medo Tr., P.E.A. ... Dc 16
Megaleen, Mat. ... Cc 15
Mehatich, P.E.A. ... Fd 15
Meirings Poort, pass, C.C. ... Bf 9
Melanie, Mt., O.F.S. ... Cc 10
Melkbosch, C.C. ... Bc
Melkbosh Point, C.C. ... Bd 7
Melmoth, Zul. ... Ee 10
Melville, C.C. ... Cf
Memba Bay, P.E.A. ... Gd 3
Memonda, B.C.A. ... Cb
Mempe, S.A.R. ... Fc 13
Mequating, O.F.S. ... Ac 10
Mere Metr, C.F.S. ... Bc 16
Merengi, P.E.A. ... Ec 16
Merone, O.F.S. ... Cc 10
Morone, O.F.S. ... Fb 7
Messaua, O.F.S. ... Dc 16
Meru, Lake, C.A. ... Bb 16
Mesa Mt., C.C. ... Ec 10
Mesire Shiramba, B.C.A. ... Cc 16
Mesumbe, B.C.A. ... Cc 16
Mewittville, S.A.R. ... 12a
Mfunga, C.F.S. ... Bb 16
Mgost, B.C.A. ... Bb 16
Mhlambve, C.C. ... Bg 10
Middelang, C.C. ... Cb 12
Middelburg, S.A.R. ... Ed 13
Middelburg, dist., C.C. ... Dd
Middelburg Road Sta., C.C. ... Ed 9
Middel Roggeveld, The, C.C. ... Ec 8
Middleveld, The, C.C. ... Dc 7
Middleton, C.C. ... Ec 9
Midiram, P.E.A. ... Ec 16
Mikati, R., P.E.A. ... Dd 16
Milanli Mt., Ny. ... Dd 16

Milk, C.C. ... Ed 7
Milkbosch Point, C.C. ... Ab 8
Milk R., C.C. ... De 9
Milkwood, C.C. ... Fg 8
Mill R., O.F.S. ... Gb 7
Mill River, town, C.C. ... Bf 9
Millwood, C.C. ... Bf 9
Mimosa, C.C. ... Db 7
Minenga, B.C.A. ... Cb 15
Mingar, S.A.R. ... Cc 12
Minula, C.F.S. ... Bb 16
Mirambo, C.F.S. ... Bh 16
Miranja, P.E.A. ... Cd 16
Misaaka, dist., B.C.A. ... Cc 16
Mitchell's Pass, C.C. ... Df 8
Mitete, P.E.A. ... Bc 16
Mitondo, Ny. ... Dc 16
Mitsi Bokidko, Bech. ... Bc 15
Mtula, P.E.A. ... Bc 16
Miyui, P.E.A. ... Cd 16
Mjema, C.F.S. ... Bc 16
Mkalwith, P.E.A. ... Cc 16
Mkalwini, in., P.E.A. ... Dc 16
Mkande, C.F.S. ... Bb 16
Mkanyo, P.E.A. ... Dc 15
Mkapo, P.E.A. ... Dc 16
Mkayigua, P.E.A. ... Dc 16
Mkewo, B.C.A. ... Cc 16
Mkinga Mts., C.F.S. ... Bb 16
Mkonolo, P.E.A. ... Dc 16
Mkongogotto, C.F.S. ... Ac 16
Mkonoma, P.E.A. ... De 16
Mkpoka, P.E.A. ... Dc 16
Mkorona, P.E.A. ... Dc 16
Mkota, P.E.A. ... Ec 16
Mkubare, R., P.E.A. ... Dc 16
Mkufi, P.E.A. ... Dc 16
Mkwali R., Bas. ... Bd 10
Mlala, P.E.A. ... Ed 16
Mlawilo, P.E.A. ... De 16
Mlangata, B.C.A. ... Cb 16
Mlinnga, B.C.A. ... Bb 16
Mlungu, B.C.A. ... Be 16
Mnabo, P.E.A. ... Dc 16
Moamba, B.C.A. ... Cc 16
Moami, B.C.A. ... Bc 16
Moani (R. Kafue), B.C.A. ... Bd 16
Moena (R. Zambesi), B.C.A. ... Bd 16
Moashemba, B.C.A. ... Bd 16
Mohomba L. (R. Zambesi), C.A. ... Ab 15
Mocinune, Bech. ... Cc 4
Mochuri, Bech. ... Bc 12
Modder Fontein, C.C. ... Cc 8
Modder Fontein Pan, Bech. ... Bb 9
Modder Fontein Pan, Klein, C.C. ... Bc 9
Modderport, O.F.S. ... Gb 9
Modder R., S.A. ... Fb 9
Modder Vlei, Great, C.C. ... Dc 7
Modimo, Bech. ... Ea 7
Modinnule, S.A.R. ... De 13
Modjadje, S.A.R. ... Fb 13
Moero, P.E.A. ... De 16
Moero, Lake, C.A. ... Bc 16
Moginqinle, P.E.A. ... Kd 16
Mogonono, Bech. ... Ac 12
Mogukruniba, P.E.A. ... Dd 16
Mohales Hoek, Bas. ... Ac 10
Mohmgura, P.E.A. ... Dd 4
Moi Duaula, P.W.A. ... Ad 4
Mollon, S.A.R. ... Be 12
Moilos, S.A.R. ... Bc 13
Moine Lema, B.C.A. ... Cc 16
Moigwe Matsa, Ny. ... Cb 16
Mojenga, P.W.A. ... Aa 4
Mokanubo Bay, P.E.A. ... Ed 16
Mokanda, Ny. ... Cc 16
Mokoil Well, Bech. ... Bd 9
Mokokongani, Bech. ... Bc 15
Mokopon, P.ech. ... Bc 4
Mokosso, C.F.S. ... Bc 16
Mokotani, Bech. ... Cc 15
Mokuana, Bas. ... Bc 10
Mokrunib, B.C.A. ... Bc 16
Mokurdula, B.C.A. ... Cb 16
Molamma, Bech. ... Ba 7
Molapo, Bas. ... Bc 10
Molei, B.C.A. ... Bd 16
Molela, B.C.A. ... Bd 16
Molen R., O.F.S. ... Cc 10
Molepolole, Bech. ... Ac 12
Mologotol, Bech. ... Cb 16
Molisani Tr., Bas. ... Fb 7
Moloka, P.E.A. ... Fc 15
Mokokoso, P.E.A. ... Ec 16
Molombe R., Mat. ... Cc 16
Molono, M., P.W.A. ... Aa 4
Molopo R., S.A. ... Cg 3
Molotsi Gold Field, S.A.R. ... Fb 13
Molototsi R., S.A.R. ... Fb 13
Moltene, C.C. ... Fd 10
Moltke, Mt., Mash. ... Fb 15
Molugwi, R., P.E.A. ... Dd 16
Mona L., P.E.A. ... Dd 16

Momba, Sw. ... Ea 10
Mombazi R., B.C.A. ... Cb 16
Mombern, Ny. ... Cc 16
Monbern (Stevenson Road), Ny. ... Cb 16
Monbi Gold Field, Mash. ... Dc 15
Mombo, B.C.A. ... Cb 16
Mona Bamba, C.F.S. ... Ab 16
Mona Kamle, C.F.S. ... Ab 16
Mona Kioni, C.F.S. ... Ab 16
Mona Maugi, C.F.S. ... Ab 16
Mona Mucanda, C.F.S. ... Ab 16
Mona Tenda, C.F.S. ... Ab 16
Mongh, P.E.A. ... Cd 16
Monika, Bas. ... Ad 10
Moniga Mt., P.E.A. ... Dd 16
Monjehodi, S.A.R. ... Eb 13
Monnaboli R., Bas. ... Cd 10
Monuque, Bech. ... Bb 12
Montagne, C.C. ... Ef 8
Montague Pass, C.C. ... Bf 9
Montai, C.C. ... Be 10
Mont aux Sources, S.A. ... Gb 10
Montepes Bay, P.E.A. ... Ec 16
Montsioa, Bech. ... Ad 13
Mouze, B.C.A. ... Cb 15
Moodie's Berg, S.A.R. ... Fd 13
Mpoerto, B.C.A. ... Be 10
Moodie's Gold Field, S.A.R. ... Gd 13
Mootfontein, S.A.R. ... Dd 13
Mooi Loop Spruit, S.A.R. ... Cc 13
Mooiplaats, S.A.R. ... Dd 13
Mooi R., G.S.W.A. ... Gb 10
Mooi R., Nat. ... Dd 10
Mooi R., S.A.R. ... Fa 7
Moora Dr., S.A.R. ... Ec 13
Moordenaars Karroo, ...
Moordenaars R., C.C. ... Be 9
Moos R., S.A.R. ... Ed 13
Mopea, P.E.A. ... Dd 16
Moramnbala Mt., P.E.A. ... Dd 16
Morati Mts., Bech. ... Ec 12
Morapae, B.C.A. ... Bc 16
Morchundi, Nat. ... Dc 4
Moremi, Bech. ... Ad 15
Morgan, Cape, C.C. ... Bg 10
Morgenzon, S.A.R. ... Cd 13
Morija, Bas. ... Ad 10
Morley, C.C. ... Bf 10
Moroka, dist., O.F.S. ... Fb 9
Morokweng, Bech. ... Bc 4
Morramnan, Bech. ... Cd 15
Morris Point, C.C. ... Cc 7
Mortimer, C.C. ... Ec 9
Morumban Satvueat (R. Zambesi), P.E.A. ... Cd 16
Morungabubara, B.C.A. ... Cc 16
Mosanga I. (R. Zambesi), S.A. ... Cb 15
Mosanco I. (R. Zambesi), ... Ad 16
Moschwane, Bech. ... Ac 12
Mosego, S.A.R. ... Ec 13
Mosgille, S.A.R. ... Dd 13
Moshoro, B.C.A. ... Be 16
Mosbanenga, Bech. ... Cb 4
Mosing, Bech. ... Bd 15
Mosita, Bech. ... Ac 12
Mosit-ode, S.A.R. ... Gc 13
Mosael Bay, C.C. ... Bg 9
Mosmas, P.E.A. ... Cd 16
Mosnpia, R., Bech. ... Ac 12
Motal Mt., Bas. ... Bd 10
Motala, S.A.R. ... Eb 13
Motel, R., S.A.R. ... Fa 13
Motenawa Hill, P.E.A. ... Ea 16
Motepnesi, R., P.E.A. ... Dc 16
Mosito, Bech. ... Da 7
Motlatsa, Bech. ... Bd 15
Motlohotlo, Bech. ... Bd 15
Motloshanena, Bech. ... Cb 4
Morsitlana, Bech. ... Ac 16
Mosita, Bech. ... Au 12
Mosset Bay, C.C. ... Bg 9
Mount Ayliff, C.C. ... Gc 10
Mount Coke, town, C.C. ... Gc 9
Mount currue, dist., C.C. ... Cc 10
Mount Darwin Gold Field, Mash. ... Eb 15
Mount Fletcher, C.C. ... Bc 10
Mount Frere, C.C. ... Bc 10
Mount Point, C.C. ... Gd 7
Mount Stewart, C.C. ... Bf 7
Morini, P.E.A. ... Dd 16
Mowila R., P.E.A. ... Bb 16
Moyara, B.C.A. ... Bb 16
Moyeni, Bas. ... Ac 10
Mozambique Channel, E.A. ... Gf 3
Mozambique, prov., P.E.A. ... Dc 16
Mozia I. (R. Zambesi), ... Bd 16
Mozinga I. (R. Zambesi), ... Cd 16
Mozuma, B.C.A. ... Cd 16
Mpako R., C.C. ... Gd 7
Mpala, C.F.S. ... Dc 8
10

Mpalera I. (R. Zambesi) ... Bb 15
Mpambe I. (Lako Nyasa), Ny. ... Cc 16
Mpanda, P.E.A. ... De 16
Mpandaji, P.E.A. ... Ec 16
Mpande, C.F.S. ... Bb 16
Mpande I. (R. Zambesi), S.A. ... Cb 15
Mparawe, B.C.A. ... Be 16
Mpasa, Ny. ... Cb 16
Mpaschele, S.A.R. ... Dc 12
Mpashi, P.W.A. ... Aa 4
Mpassa, P.E.A. ... Dd 16
Mpassa (R. Shire), P.E.A. ... Bd 16
Mpata, B.C.A. ... Bd 16
Mpelembe, P.E.A. ... Dc 16
Mpemba, Ny. ... Cc 16
Mpembe, Ny. ... Cc 16
Mpembe, P.E.A. ... Cd 16
Mpeseni, B.C.A. ... Ed 3
Mpile Nek, C.C. ... Be 10
Mpinti, B.C.A. ... Be 16
Mpipi, C.F.S. ... Bb 16
Mpito, Bas. ... Be 10
Mponda (R. Rovuma), P.E.A. ... Fd 3
Mponda (R. Shire), Ny. ... Cc 16
Mpondwe, B.C.A. ... Bb 16
Mpundu, Ny. ... Cc 16
Mrianwendo, P.E.A. ... Dd 16
Mrila, P.E.A. ... Dd 16
Msain, R., P.E.A. ... Dc 16
Msenza, P.E.A. ... Cd 16
Msinji Valley, P.E.A. ... Bc 16
M'Siri, C.F.S. ... Bc 16
Msiri's Kingdom, C.F.S. ... Bb 16
Msea, C.F.S. ... Bb 16
Msowe, Mash. ... Ec 15
Msakilua, P.E.A. ... Fd 15
Msava, P.E.A. ... Dd 16
Mswilo, Ny. ... Cb 16
Mtanalabare, B.C.A. ... Bd 16
Mtarika (R. Lujenda), P.E.A. ... Dc 16
Mtarika (R. Rovuma), ...
Mti-bika, Mash. ... Ed 15
Mtegazi, Mt., P.E.A. ... Be 16
Mtemhunje, G.S.W.A. ... Ba 4
Mtengult, P.E.A. ... Cc 16
Mtima, P.E.A. ... Dc 16
Mtomin, Mt., P.E.A. ... Cc 16
Mtoza, P.E.A. ... Dc 16
Mtunde, I., P.E.A. ... Ec 16
Mualin, P.E.A. ... Dc 16
Muamba Sambamba, C.C. ... Cc 16
Muazi-Agola, C.F.S. ... Ac 16
Munshike, C.F.S. ... Bc 16
Munsi, R., P.E.A. ... Fc 15
Mueneta, P.E.A. ... Ed 12
Muda, R., P.E.A. ... 3a
Muaden, Nat. ... Cc 16
Mudge, Point, C.C. ... Dg 8
Mudia, P.E.A. ... Dc 16
Mudibug, S.A.R. ... Ea 7
Mudichbi, P.E.A. ... 3a
Mudzi, R., Mash. ... Fb 15
Muembe, Tr., P.E.A. ... Dd 16
Muene Auta, C.F.S. ... Bc 16
Muene Kula, C.F.S. ... Ac 16
Muense, C.F.S. ... Ab 16
Mufa R., P.E.A. ... Cd 16
Mufukushi, R., N.Z. ... Bc 16
Mugete, P.E.A. ... Ec 16
Mugube Magalo, W., Bech. ... Bb 4
Muiden, Klein, S.A.R. ... Ed 13
Mui Gallulua, B.C.A. ... Bb 16
Muira, R., P.E.A. ... Dc 16
Muiri, C.F.S. ... Ac 16
Muiza, B.C.A. ... Cc 16
Mukishond R., C.C. ... 3a
Muizhoekberg, The, C.C. ... Ce 8
Mukakanbo, C.F.S. ... Bb 16
Mukana, C.F.S. ... Ab 16
Mukunguru, S.A.R. ... Cc 16
Mukwaru, Ny. ... Cc 16
Mulanda, B.C.A. ... Bc 16
Mulandi, B.C.A. ... Ac 16
Mulders Vlei Junction, C.C. ... Cf 8
Mulua, P.E.A. ... Dc 16
Mumba, C.F.S. ... Bc 16
Mumbeje, R., P.E.A. ... Bc 16
Mumpanta Mt., B.C.A. ... Bc 16
Munkula, C.F.S. ... Bb 16
Munsa, C.F.S. ... Bb 16
Mungatu, Ny. ... Cb 16
Muniham, P.E.A. ... Fd 16
Muno Vuito, B.C.A. ... Cc 16
Munsbatext, C.F.S. ... Bc 16
Muongo, M., P.W.A. ... Aa 4
Muorango, Ny. ... Cd 16
Mural Mts., S.A.R. ... Cb 12
Murchison, S.A.R. ... Fb 13
Murchison Falls (R. Shire), Ny. ... Bc 3

Name	Sq.	Pg.
Murchison Range, S.A.R.	F b	13
Muriçane, P.E.A.	F c	15
Muroa, P.E.A.	C d	16
Muro Ashinio, B.C.A.	C c	16
Murraysburg, C.C.	C d	9
Murumbu, B.C.A.	C c	16
Musau, S.A.R.	D b	12
Musoba, B.C.A.	C c	16
Mushena, P.E.A.	C d	16
Mushinga, C.F.S.	A c	16
Musananngoe R., P.E.A.	F a	16
Musso Kantanda, P.W.A.	A c	16
Musula, C.F.S.	A c	16
Mutangwa R., P.E.A.	F b	15
Mutipa's, B.C.A.	C c	16
Mutsi R., S.A.R.	C c	12
Mutua R., N.Z.	E b	15
Mutukuta, B.C.A.	C c	16
Muxinga Mts., C.A.	C c	16
Muzigagnva, Mash.	E c	15
Mvaunbi Bay, P.E.A.	E c	16
Mvoonu, Tong.	D c	4
Mwambi, P.E.A.	G d	3
Mwemba I. (R. Zambesi), S.A.	C b	15
Mwembe, P.E.A.	D c	16
Mweru, Lake, C.A.	D c	3
Mwiganua, P.E.A.	D c	16
Mwojin, P.E.A.	D c	16
Myaka, S.A.R.	D c	12
Myafontein, C.C.	C c	9
Mzeuza, P.E.A.	E a	15
N		
NAAUW POORT JUNCTION, C.C.	E d	9
Naauwte Vlei, C.C.	F c	8
Nabalals, G.S.W.A.	D b	7
Nabna, G.S.W.A.	A c	4
Nabis, G.S.W.A.	A c	4
Naboomfontein, S.A.R.	C c	12
Nagatatolio, Bech.	B c	15
Nagaloe, P.E.A.	D d	16
Nahange, P.E.A.	F c	15
Nahanyawa, P.E.A.	D d	16
Nahhambe, C.C.	F d	7
Nahorla, P.E.A.	D d	16
Nakabcle Falls (R. Zambesi), S.A.	B d	10
Nakachinto, R., B.C.A.	E d	16
Naka Gold Field, Mat.	E d	16
Nakhusha, P.E.A.	E c	16
Nako, R., P.E.A.	F a	15
Nakwa, R., P.E.A.	D c	16
Natole, B.C.A.	A d	16
Namakau, I., B.C.A.	A d	16
Namalongo, P.E.A.	E d	16
Namaquauand, Great, G.S.W.A.	A c	
Nauuaqualond, Little, C.C.	B b	8
Namara, P.E.A.	D c	10
Namashill, P.E.A.	D b	7
Nanpasusseu, I. (R. Zambesi), S.A.	E a	15
Namboromo, B.C.A.	B c	16
Nambwe Cataract (R. Zambesi), S.A.	A d	16
Nameta, B.C.A.	A d	10
Namiburi, P.E.A.	D d	16
Namkuna, P.E.A.	D c	16
Namkwitta, P.E.A.	D d	16
Namluga, P.E.A.	D d	16
Nazalokoko, P.E.A.	E d	16
Nauurub, C.C.	B c	8
Namtusi, P.E.A.	D c	16
Namuli Peaks, P.E.A.	F c	3
Namrania, P.E.A.	D d	16
Nana, C.C.	A c	8
Nana Kandundu, B.C.A.	C d	3
Nanebis, G.S.W.A.	B a	7
Nangana, G.S.W.A.	B a	4
Nangana, P.E.A.	D d	16
Namiasi, P.E.A.	D d	16
Nantapa, P.E.A.	D d	16
Naochalieb, G.S.W.A.	A c	4
Nnoxabeli, G.S.W.A.	B a	4
Nape, C.C.	B c	7
Napier, C.C.	D g	8
Napusa, P.E.A.	D c	16
Narnab, G.S.W.A.	B a	7
Nardouw, C.C.	E d	7
Naresie, C.C.	A c	8
Narlole, B.C.A.	A d	16
Naros, G.S.W.A.	A c	4
Narugas, G.S.W.A.	B a	7
Narnhus R., S.W.A.	A b	4
Nans Nass Point, C.C.	A b	8
Natal, S.A.	D d	10
Natal, Port, Nat.	E d	10
Natal Spruit, S.A.R.	D d	13a
Nata R., Mat.	C c	15
Natchiwa, P.E.A.	C c	16
Nateuabe, P.E.A.	D d	16
Nauko, B.C.A.	B d	16
Nauwte Vlei, C.C.	C c	9
Nawaruma, P.E.A.	D d	16
Nazareth, Nat.	D b	10
Ncaulua, P.E.A.	D c	16
Ncamana, Tong.	F b	10
Nchine, P.E.A.	D c	16
Nchokotsa, Bech.	B d	15
Ndari, B.C.A.	C b	16
Ndarina, P.E.A.	D c	16
Ndava, G.S.W.A.	B a	4
Ndonda, Ny.	C c	16
Ndoomba, Tong.	E d	12
Nebulu, P.E.A.	D d	16
Nel, S.A.R.	D d	12
Nels Poort, C.C.	B c	9
Nelsville, S.A.R.	B d	13
Nena, Bas.	F b	7
Neu Halle, S.A.R.	C c	12
Neves Ferreira, P.E.A.		3A
New Amsterdam, S.A.R.	D d	12
New Bethesda, C.C.	D d	9
Newcastle, C.C.	G f	9
Newcastle, Nat.	C b	10
Newenale, co., Nat.	C c	10
New Denmark, S.A.R.	E c	13
Newdigate, Fort, S.A.R.	D c	12
New Glasgow, Nat.	E d	10
New Halle, S.A.R.	D d	12
New Pass, C.C.	B c	10
New Republic, The, S.A.R.	D b	10
New Scotland, S.A.R.	D d	12
Newtondal, C.C.	F d	7
Newton Peak, C.C.	B c	10
Newtonvile, Nat.	C c	10
Newtondale, C.C.	F f	7
New Year I., C.C.	P f	9
New Year's R., O.F.S.	G b	7
Ngabisane, Bech.	B d	13
Ngambo, R., P.E.A.	D c	16
Nguoi, Lake, Bech.	A d	15
Ngoanessi R., S.A.R.	D b	12
Ngombe, Ny.	D c	16
Ngunga, Bech.	D a	7
Ngwa Hill, Bech.	B c	15
Nhameassunsuva, Mash.	F b	15
Niamwaits, P.W.A.	A c	10
Nickerk, C.C.	D d	12
Niemands, C.C.	F f	9
Nieuwveld, The, C.C.	C c	7
Nieuwveld Range, C.C.	B c	9
Nifalu, B.C.A.	A d	16
Nihegeli R., P.E.A.	E c	16
Nilonaa, P.E.A.	D d	16
Nikungu, P.E.A.	D c	16
Nikutu, P.E.A.	D c	16
Niusi, P.E.A.	C c	
Njoko, R., B.C.A.	A d	16
Nkandhla, dist., Zul.	D c	10
Nkoebe's, Bas.	A c	10
Nkuka's Kraal, C.C.	C f	10
Nkuuakwe, C.C.	A g	16
Nkumba, Ny.	D c	16
Nojo, P.W.A.	A c	16
Nokauua R., Bech.	D b	7
Nolloth, Port, C.C.	A b	8
Noma Fall, G.S.W.A.	B a	4
Nondwans, P.E.A.	E c	12
Nonjes Poort, C.C.	D d	8
Nonkonyani, C.C.	C e	10
Nonuwe R., P.E.A.	D d	16
Noodsberg, The, Nat.	D d	10
Norden, S.A.R.	D d	12
Northampton, Fort, S.A.R.	D c	16
North Sand Bluff, Nat.	D f	10
Northumberland Point, C.C.	E g	8
Norubi, C.C.	C u	9
Nosob R., G.S.W.A.	B f	5
Nosop R., Bech.	B b	6
Nosop, W. Black, S.A.	A b	4
Nosop, W. White, G.S.W.A.	A b	4
Noosi Ve., Mat.	C f	8
Notong, C.C.	H c	7
Nottingham, Fort, Nat.	G b	7
Notwani, R., Bech.	C b	4
Noup Plateau, S.W.A.	D a	8
Nqusi, C.C.	B c	4
Nquatsha's, Bas.	B c	10
Nquta Mts., S.A.R.	D c	10
Nrugi Mt., P.E.A.	E b	12
Nsoha, Zul.	E c	12
Nsuto R., Great, Sw.	D d	12
N'Teuke, O.F.S.	B c	10
Ntunda, P.E.A.	D d	16
Ntwara, Ny.	C e	16
Ntwe-Ntwe Salt Pan, Bech.	B d	15
Nuaunetsi, R., Mat.	E d	15
Nugumes, G.S.W.A.	A a	7
Nugosis, G.S.W.A.	B a	7
Nujania, Bech.	C e	3
Numas, C.C.	B c	4
Numees, C.C.	A a	8
Nutzi, R., C.C.	C g	9
Nwauetsi, R., P.E.A.	H c	13
Nyadimba, I. (R. Zambesi), S.A.		
Nyakoba, P.E.A.	F a	15
Nyamatarara, R., P.E.A.	C d	16
Nyamounga, P.E.A.	C d	16
Nyampanga, I. (R. Zambesi), S.A.	C d	16
Nyampunga I. (R. Zambesi), S.A.	B d	16
Nyango, P.E.A.	D b	15
Nyaondwe, P.E.A.	D d	16
Nyassa, lake, C.A.	F a	15
Nyassaland, B.C.A.	E d	3
Nyawos Hill, S.W.A.	C c	16
Nyawoskop, The, S.A.R.	D d	12
Nyema Kapemba, B.C.A.	E b	10
Nyimbu, Ny.	C b	16
Nyl R., S.A.R.	C c	16
Nyl R., S.A.R.	C b	12
Nylstroom, S.A.R.	D b	13
Nylstroom R., S.A.R.	D e	13
Nyl Vlei, S.A.R.	D c	13
O		
OANGWA, R., S.Z.	C d	16
Oas, G.S.W.A.	A b	4
Obere Zak R., C.C.	C b	7
Oboop, C.C.	D a	8
Odendabl, O.F.S.	C c	4
Odonga, G.S.W.A.	B c	3
Odzi R., Mash.	P f	9
Oertel, S.A.R.	D c	12
Oesterhuys, S.A.R.	C c	12
Oham, S.A.R.	D d	12
Ohambahando, G.S.W.A.	A b	4
Okahandya, G.S.W.A.	A b	4
Okanahati, G.S.W.A.	A a	4
Okanbombo, G.S.W.A.	A a	4
Okavango R., B.C.A.	B c	3
Okavarona, G.S.W.A.	A b	4
Okonokwa, Bech.	B b	4
Old Buntingville, C.C.	C f	10
Old Tsolo, C.C.	B f	10
Olifant Barg, The, C.C.	B d	7
Olifant, Port, S.A.R.	B c	9
Oliphantshoek Pt., C.C.	C g	9
Oliphants Fontein, O.F.S.	D a	9
Oliphants Mts., C.C.	D c	8
Oliphant R., S.A.R.	E d	13
Oliphants R. (Carnarvon), C.C.	C c	9
Oliphants R. (Ladismith), C.C.	D c	8
Oliphants R. (Jan William), C.C.	D c	8
Oliphants R., Klein, S.A.R.	E d	13
Oliphants Vlei, C.C.	F c	8
Olifants Vley, S.A.R.	C c	9
Oliphants Vlei R., C.C.	F c	8
Omaruulpa Spuriko, Sw.	A b	4
Onaramba R., S.W.A.	B a	4
Omba Omenzi, G.S.W.A.	A a	4
Ombongo, G.S.W.A.	A a	4
Omhongo Tr., S.W.A.	A a	4
Omhoronbonga, G.S.W.A.	A b	4
Omdraai, Bech.	B b	4
Oneva, G.S.W.A.	A a	4
Omurumbourn, G.S.W.A.	A b	4
Omohira, P.W.A.	A a	4
Onandaya, G.S.W.A.	A a	4
Onehas, G.S.W.A.	A c	4
Onlerste Doorns, C.C.	E c	8
Onderveld, The, Bech.	C d	15
Ongar R., C.C.	C d	9
Ongelnks R., C.C.	D e	8
Ongeluk, C.C.	D b	7
Ongeluk R., C.C.	D c	7
Ongeluk's Nek, S.A.	B e	10
Onkoro Okavupa, G.S.W.A.	A a	4
Ongova Tr., S.Z.	B a	4
Onsila R., G.S.W.A.	A a	4
Ontilimani, C.C.	A b	9
Oobiep, C.C.	A b	8
Oouay, R., S.Z.	B d	15
Oori R., P.E.A.	E b	12
Oorlogs Kloof R., C.C.	B b	8
Orange R., S.A.	A b	8
Orange Free State, S.A.	C b	9
Orange R., Mouth of the, C.C.	A a	8
Orange River Sta., C.C.	D b	9
Origstad, S.A.R.	E c	13
Oriogs R. (Clanwilliam), C.C.	B c	7
Oriogs R. (Colesberg), C.C.	E c	3
Oroab, Bech.	B c	4
Orob, The, C.C.	C a	8
Oru Point, Tong.	F a	10
Oruthe, G.S.W.A.	A a	4
Os Berg, The, C.C.	B c	8
Oscar, Nat.	D c	10
Osse Spruit, O.F.S.	E b	7
Othello, Nat.	D c	10
Otinati, Nat.	E d	10
Otjijika Mts., G.S.W.A.	A a	4
Otjirandu, C.C.	B d	7
Otter Pan, C.C.	B c	9
Ottoshoop, S.A.R.	A d	13
Otvita, G.S.W.A.	A a	4
Otyirc, G.S.W.A.	A b	4
Otyikeko, G.S.W.A.	A b	4
Otyikeko, G.S.W.A.	A a	4
Otyimbinde, G.S.W.A.	B b	4
Otyimbindo, Wady, G.S.W.A.	A b	4
Ouvimbinka, G.S.W.A.	A b	4
Ovionakoyo, G.S.W.A.	A a	4
Otyire, G.S.W.A.	A b	4
Otyisoma, G.S.W.A.	A b	4
Oljonakala Berg, The, G.S.W.A.	A a	4
Otyosan, G.S.W.A.	A b	4
Oxyozondyupa, G.S.W.A.	A b	4
Oubeep Cove, C.C.	A b	8
Ouchas, S.A.R.	A c	4
Ouddshoorn, C.C.	B f	9
Oup R., Bech.	B b	4
Ours R., C.C.	C c	9
Ousema, G.S.W.A.	A b	4
Ouenigna, Mts., C.C.	B f	9
Ovambo Tr., G.S.W.A.	B c	3
Ovatyimba Tr., G.S.W.A.	A b	4
Overtoun, Nat.	D c	10
Ozire, G.S.W.A.	A b	4
Oumbeyakauha, G.S.W.A.	A b	4
P		
PAARDE BERG, THE, C.C.	C f	8
Paarde Kraal, C.C.	G d	9
Paarde Kraal, O.F.S.	A b	10
Paardepoort, pass, C.C.	B f	9
Paard Fontein, C.C.	E d	7
Paarl, C.C.	C f	8
Paauw Pan, C.C.	B c	9
Pacaltsdorp, C.C.	B f	9
Pack tiu Nek, S.A.	B c	10
Padclis, C.C.	B c	10
Paulrone, Cape, C.C.	F g	9
Paerzylmloo R., S.A.R.	D b	13
Pafuri, S.A.R.	D c	13
Pafuri R., S.A.R.	G a	13
Pagadi, P.E.A.	F d	15
Pabla, Mat.	D d	15
Pajodzi R., P.E.A.	F a	15
Pakaiti, Nat.	D c	10
Pakalimpama, B.C.A.	B c	16
Pakambwera, Ny.	C c	16
Pakariro, O.F.S.	B c	10
Pakaundi, dist., C.C.	C c	16
Pakwe R., Mat.	C b	12
Palala R., Great, S.A.R.	C b	12
Palani, Mt. M., S.A.	C b	7
Palapye, Bech.	C c	15
Palmiet R., C.C.	C g	8
Palnerton, C.C.	C c	16
Pamalouliwe Lake, Ny.	C c	16
Pambala, Ny.	C c	16
Panbete, B.C.A.	C b	16
Pampoen Pan, C.C.	C b	9
Pampoen Poort, C.C.	D d	9
Pandu-ma-Tenka, Bech.	B c	15
Pando, C.F.S.	B c	16
Punga, Mt., P.E.A.	F c	15
Pangani, Itas, P.E.A.	E c	16
Pangara, P.E.A.	D d	16
Pangola, P.E.A.	E d	13
Pangnona, Tr., C.A.	F c	15
Pamaure, C.C.	D c	16
Pantula, Ny.	E b	16
Panyame Mt., P.E.A.	E b	15
Papauiol, C.C.	D b	7
Papendorf, C.C.	A a	8
Papkuil, C.C.	D b	7
Parapato, P.E.A.	F c	15
Parijs, O.F.S.	D d	13
Passes, P.E.A.	C d	16
Patehi, Bas.	C c	10
Paternoster Point, Great, C.C.	B e	8
Paterson, C.C.	E f	9
Patrys Berg, The, C.C.	C b	8
Pattuerson, C.C.	A g	10
Patuni, Bech.	B e	4
Pauho, Ny.	C c	16
Paul Pieters Dorp, S.A.R.	D c	4
Pazanzan, P.E.A.	D c	16
Peacock Roads, C.C.	B c	9
Pearson, Fort, Nat.	E d	10
Pearston, C.C.	E c	9
Peddic, C.C.	F f	9
Pedros Kloof, C.C.	B c	7
Peelton, C.C.	G c	9
Pekawi, P.E.A.	E b	16
Pekawi, Ras, P.E.A.	E c	16

Pella, C.C. ... D b 8
Pella, S.A.R. ... E d 13
Pella, Little, C.C. ... C b 8
Pemba Bay, P.E.A. ... E c 16
Penguin Rock, C.C. ... A b 8
Penguins Nek, Sw. ... E b 10
Pennings Drift, Bech. ... C c 12
Perie, C.C. ... G e 9
Perigongi, P.E.A. ... C d 16
Pesimba, P.E.A. ... D e 16
Petrusberg, O.F.S. ... E b 9
Petrusville, C.C. ... D e 9
Philadelphia, C.C. ... G f 8
Philippolis, O.F.S. ... E c 9
Phillipstown, C.C. ... D e 9
Pietermaritzburg, Nat. ... D d 16
Pietersburg, S.A.R. ... E b 13
Pieters, S.A.R. ... B e 12
Piet Potgieter's Rust, S.A.R. ... D c 13
Piet Retief, S.A.R. ... F f 13
Pig's Peak, Sw. ... G d 13
Pilani, Bech. ... A e 12
Pilands Berg, S.A.R. ... C d 13
Pillar Kraal, Nat. ... C d 15
Pilgrim's Rest, S.A.R. ... F e 13
Pine, Fort, Nat. ... D e 16
Pinetown, Nat. ... D d 16
Pingwe, Ny. ... D d 16
Piosela, P.E.A. ... F d 15
Piquetberg, C.C. ... C e 8
Piquetberg Road Sta., C.C. ... C f 8
Piquet Berg, The, C.C. ... C f 8
Piru, Mt., B.C.A. ... A e 16
Piric's, S.A.R. ... D b 4
Pisangkop, The, S.A.R. ... F b 15
Pisene, P.E.A. ... D d 13
Pisini, P.E.A. ... E c 12
Pitlanganyane, Bech. ... A e 12
Pitsan, S.A.R. ... D c 12
Pitsanie, Bech. ... A e 12
Platberg, The, S.A. ... E a 9
Plat R., S.A.R. ... D c 13
Plessis R., C.C. ... C f 9
Plettenberg, C.C. ... C g 9
Plettenberg Bay, C.C. ... C g 9
Pniel, C.C. ... D a 9
Pocho, Bas. ... B c 10
Pocho's Peak, O.F.S. ... B e 10
Poklonos Kop, The, S.A.R. ... D b 12
Poko, Ny. ... G c 13
Pokollo Cataract, (R. Ochompo), B.C.A. ... A e 16
Pokuteke R., Mash. ... E e 15
Pokwani, Bech. ... E a 7
Polfontein, S.A.R. ... A d 12
Polonia, S.A.R. ... C d 13
Pomfia Bay, P.E.A. ... G d 3
Pomeroy, S.A.R. ... D e 10
Pompam Pan, C.C. ... D b 7
Pondoland, C.C. ... C f 10
Pongola R. (Utrecht), S.A.R. ... D b 10
Pongola R. (Waterberg), S.A.R. ... C e 13
Poortbesdam, C.C. ... E b 7
Poortjes Fontein, O.F.S. ... C c 4
Pt. Alfred, C.C. ... F f 9
Port Beaufort, C.C. ... E g 8
Port Elizabeth, C.C. ... E f 9
Porterville, C.C. ... C c 8
Port Herald, Ny. ... F e 3
Port Natal, Nat. ... E d 16
Port Nolloth, C.C. ... A b 8
Port Shepstone, Nat. ... D e 16
Portuguese East Africa, ... D d 16
Post Relief, C.C. ... F e 9
Potchefstroom, S.A.R. ... C e 9
Potfontein, C.C. ... C e 9
Potgieter (Rustenburg), S.A.R. ... A e 7
Potgieter (Rustenburg), S.A.R. ... B c 12
Potgieters Rust, S.A.R. ... C e 12
Pot R., C.C. ... B f 10
Potzdam, C.C. ... C e 9
Pram Bergen, The, C.C. ... B d 9
Pram Berg, The, C.C. ... B d 9
Prela, S.A.R. ... G a 7
Predict, S.A.R. ... C c 12
Pretoria, S.A.R. ... D d 13
Pretoria (Inset Map of), S.A.R. ... 12A
Pretoria (Heidelberg), S.A.R. ... D e 13
Pretoria (Pretoria), S.A.R. ... D e 13
Pretoria (Rustenburg), S.A.R. ... D e 13
Prieska, C.C. ... B b 9
Primeira Is., P.E.A. ... F d 16
Prince Albert, C.C. ... F f 8
Prince Albert, dist., C.C. ... E f 9
Prince Albert Road Sta., ... F e 8
Prince Alfred, C.C. ... D f 8

Prince Alfred's Pass, C.C. ... C f 9
Prinslo, S.A.R. ... D d 13
Prinsloo (Pretoria), S.A.R. ... C e 12
Prinsloo (Pretoria), S.A.R. ... C d 12
Priors, O.F.S. ... E c 9
Proces Fontein, C.C. ... D c 7
Providential Gorge, Mash. ... E d 15
Puffadder, C.C. ... B b 7
Pugat's, P.E.A. ... F e 13
Paln Mid, P.E.A. ... F d 13
Pungwe R., P.E.A. ... E c 3
Puttera Kraal, C.C. ... F d 9

Q

Quani, Bech. ... B c 15
Quaggas Fontein, C.C. ... C e 7
Quaggas Puts, C.C. ... D e 8
Qualimbata, Bech. ... B c 15
Quamacee, C.C. ... A f 10
Quatlatsila, Mash. ... D d 15
Qudeni Mts, Zul. ... D e 10
Quedlingburg, S.A.R. ... E b 13
Queenstown, C.C. ... F d 9
Querimba Is., P.E.A. ... E e 16
Quikura Falls (Luapula R.), C.F.S. ... B b 16
Quilimane, P.K.A. ... D d 16
Quilimane R., P.E.A. ... D d 16
Quinzungu I., P.E.A. ... D d 16
Qumbo, C.C. ... B f 10
Quoin Point, C.C. ... D g 8
Quorn R., C.C. ... G d 7
Quthing, see Nuthing.

R

Radloff, C.C. ... E b 7
Rahame Pass, Bas. ... B e 10
Rawageep, S.A.R. ... E b 13
Ramah, C.C. ... D b 9
Ramaquaban R., Tati ... C d 15
Ramatlabama, Bech. ... A d 13
Ramatlabama Pool, Bech. ... C e 4
Rame Head, C.C. ... C f 10
Ramkwa Samassin, ...
Ramoutska, Bech. ... B c 12
Rand Berg, The, S.A.R. ... B c 12
Rareka R., P.E.A. ... D d 16
Ras Pungani, P.E.A. ... E e 16
Ras Pekawi, P.E.A. ... E c 16
Ratabane, O.F.S. ... F b 7
Rawlinson Mt., S.A.R. ... D d 12
Raynor, C.C. ... D d 12
Real's Drift, C.C. ... C b 9
Rebanga, S.A.R. ... F b 13
Rebel, B.C.A. ... B d 16
Recife, Cape, C.C. ... E g 9
Reddersberg, O.F.S. ... F b 9
Red House, C.C. ... E f 7
Reef Point, C.C. ... G d 7
Rehoboth, G.S.W.A. ... A b 4
Reitz, O.F.S. ... B b 10
Reitzburg, O.F.S. ... C e 13
Remhsburg, C.C. ... E c 9
Rensburg, S.A.R. ... F d 13
Rennicke, S.A.R. ... F a 7
Rensburg (Ermelo), S.A.R. ... D d 13
Rensburg (Potchefstroom), S.A.R. ... F a 7
Rensburg (Rustenburg), S.A.R. ... C d 13
Retief, S.A.R. ...
Reuben, C.C. ... C d 16
Revnbwe, R., P.E.A. ... 3A
Revne, R., P.E.A. ...
Rhabetsani, Bech. ... B c 12
Ranasltuase, S.A.R. ... E c 7
Rhenoster Fontein, O.F.S. ... B c 7
Rhenosterkop Sta., C.C. ... F a 7
Rhenoster Kop, The, O.F.S. ... E d 13
Rhenoster Poort, S.A.R. ... F b 9
Rhenoster R. (Bloemfontein), O.F.S. ...
Rhenoster R. (Hopetown), C.C. ... F a 7
Rhenoster R. (Kroonstad), O.F.S. ... C c
Rhenoster R. (Pretoria), S.A.R. ... E e 8
Rhenoster R. (Sutherland), C.C. ... C c 8
Rhenoster Valley, C.C. ...
Rhodesia, B.C.A. ... B c
Ribumi Fontein, Bech. ... E e 10
Richards Bay, Zul. ... C d 9
Richmond, C.C. ... D d 16
Richmond, Nat. ... F g 9
Richmond Hills, C.C. ...

Richmond Road Sta., C.C. ... C d 16
Richterveld, The, C.C. ... A a 8
Ricketsdam, S.A.R. ... B e 12
Ridsolo R., S.A.R. ... F b 13
Riebeck, C.C. ... F f 9
Riebeekcastcel, C.C. ... C f 8
Ricker, S.A.R. ... C c 12
Riet, C.C. ... B e 7
Rietfontein, Bech. ... C f 3
Rietfontein, G.S.W.A. ... B a 7
Rietfontein, S.A.R. ... E a 7
Riet Fontein (Albert), C.C. ... E c 7
Rietfontein (Carnarvon), C.C. ... F a 8
Riet Fontein (Colesberg), C.C. ... E c 7
Riet Fontein (Great Bushman land), C.C. ... C b 7
Riet Fontein (Griqualand West), C.C. ... D b 9
Rietfontein (Richmond), C.C. ... C e 9
Riet R., Klein, C.C. ... C a 9
Riet, Port, C.C. ... C d 7
Riet Point, C.C. ... G f 9
Riet R., O.F.S. ... F b 9
Riet R., S.A.R. ... C e 13
Riet R. (Ceres), C.C. ... D e 8
Riet R. (Fraserburg), C.C. ... C e 7
Riet R. (Griqualand West), C.C. ... D b 9
Riet R., Great (Somerset East), C.C. ... E c 9
Riet R., Great (Sutherland), C.C. ... E e 8
Riet R., Little, C.C. ... F d 8
Riet Spruit, O.F.S. ... F b 9
Riet Spruit, S.A.R. ... C e 12
Riet Vlei, C.C. ... C e 10
Rikuru R., Ny. ... C c 16
Ritofil, P.E.A. ... E c 16
Riversdale, C.C. ... F g 8
Riverton, Nat. ... D d 16
Robbe Bay, C.C. ... A b 7
Robben I., C.C. ... D f 8
Robertson, C.C. ... D e 8
Robinson Pass, C.C. ... A f 9
Rode, C.C. ... C e 10
Rodewal, C.C. ... C d 7
Rodi Duinen Point, C.C. ... C d 8
Rogeveld Mts., C.C. ... E e 8
Roggeveld, The Achte, C.C. ... E d 8
Roggeveld, The Klein, C.C. ... E o 8
Roggeveld, The Middel, C.C. ... E e 8
Rohff's I. (R. Zambesi), S.A. ... A d 16
Rooikloofrontein, S.A.R. ... D d 12
Roma, Bas. ... A d 10
Roman Vloer, C.C. ... E c 8
Roodashe, Ny. ... E c 16
Roodale, C.C. ... A c 7
Rood Berg, The, C.C. ... F b 9
Roodo, P.E.A. ... D d 16
Roodepat, C.C. ... C c 8
Roodeval, C.C. ... B d 7
Rooikrantz, C.C. ... B d 9
Roode Berg (Aberdeen), The, C.C. ... D e 9
Roode Berg (Ladismith Dist.), The, C.C. ... F f 8
Roode Berg (Middelburg), The, C.C. ... E c 7
Roode Klip R., C.C. ... C f 8
Roode Rand, The, Sw. ... D e 13
Roode Vloer, Pan, C.C. ... D b 7
Rooikwal, S.A.R. ... B c 12
Rooi Berg, C.C. ... B c 8
Rooi Grond, The, S.A.R. ... A d 13
Rooi Berg, The, S.A.R. ... C c 12
Roos, S.A.R. ... F a 7
Roos R., S.A.R. ... E d 13
Rooswankal, S.A.R. ...
Rorke's Drift, Nat. ... D e 10
Rose Fontein, C.C. ... E d 9
Roseentecville, S.A.R. ...
Rosi Mopani, Bech. ... B c
Rouxville, O.F.S. ... F c 9
Rovuma Bay, P.E.A. ... D c 16
Rovuma R., E.A. ... D c 16
Rowe, S.A.R. ... F c 7
Ruatasi, P.E.A. ... D c 16
Rugani, Tong. ... E d 12
Ruggens, The Zwarte, C.C. ... D f 8
Ruimsi R., S.Z. ... E b 15
Ruigtefontein, O.F.S. ... A e 7
Ruimaire R., Mash. ... F b 15
Ruo, R., B.C.A. ... D d 16
Rupert, Mt., C.C. ... D a 9
Rusambo, Mash. ... F b 15
Rustenburg, S.A.R. ... C c 12
Ruzarwe, R., Mash. ... E c 15

S

Sabine, Tati ... C d 15
Sabi, R., Mat. ... E c 15
Sabi R., S.A.R. ... G e 13
Sabiai, Bech. ... B b 4
Sadum R., S.Z. ... B a 4
Sadya's, Mash. ... E c 15
St. Albans, C.C. ... A f 10
St. Andrews, C.C. ... E c 7
St. Andrews, Zul. ... E d 10
St. Augustine, C.C. ... G c 7
St. Augustine, C.C. ... D c 10
St. Blaise, Cape, C.C. ... E g 9
St. Croix I., C.C. ... E f 9
St. Francis Bay, C.C. ... E g 9
St. Francis, Cape, C.C. ... D g 9
St. George R., P.E.A. ... E c 12
St. Helena Bay, C.C. ... C e 8
St. James, Zul. ... E c 10
St. John's, C.C. ... G c 7
St. John's R., C.C. ... C f 10
St. Lazarus Bank, P.E.A. ... E c 16
St. Lucia Bay, S.E.A. ... F e 10
St. Lucia, Cape, Zul. ... F e 10
St. Lucia Lake, S.E.A. ... F b 10
St. Marks, C.C. ... G d 9
St. Martin, Cape, C.C. ... A d 7
St. Mary Cape, Mad. ... C g 3
St. Michaels, Nat. ... D e 10
St. Mingo Bay, C.C. ... E g 3
St. Paul's, Zul. ... E c 10
St. Peters, C.C. ... G e 9
St. Philip, Zul. ... E e 10
St. Sebastian Bay, C.C. ... E g 16
Sakatoko, Mash. ... E c 15
Sakun Mts., S.A.R. ... F e 13
Salati R., S.A.R. ... D c 12
Saldanha Bay, C.C. ... B f 8
Salem, C.C. ... F f 9
Salisbury, Mash. ... E b 15
Salisbury Gold Field, Mash. ... E b 15
Salons R., S.A.R. ... B c 12
Salt R. (Beaufort West), C.C. ...
Salt R. (Cape), C.C. ... C e 9
Salt R. (Fraserburg), C.C. ... C f 8
Salt R. (Great Bushman Land), C.C. ... F d 8
Sama R., S.A.R. ... D b 8
Saubana, Tong. ... F b 13
Sambone, Sw. ... F b 10
Sambuti R., Bech. ... D d 12
Samson's Gat, C.C. ... A c 13
Samson's Gat, C.C. ... B b 7
Sana Bashi, P.W.A. ... A e 16
Sanacan, Cape, P.K.A. ... E e 16
Sandberg, S.A.R. ... F c 13
Sand Bluff, North, Nat. ... D f 10
Sand Bluff, South, C.C. ... D f 10
Sand Flats, C.C. ... E f 9
Sandfontein, Bech. ... B b 4
Sandfontein, G.S.W.A. ... B b 7
Sanda, P.E.A. ... C d 16
Sandown Bay, C.C. ... D g 7
Sandown Point, C.C. ... A b 8
Sand R., Nat. ... G b 7
Sand R. (Vrede), O.F.S. ... E f 13
Sand R. (Winburg), O.F.S. ...
Sand R. (Waterberg), S.A.R. ... C b 12
Sand R. (Zoutpansberg), S.A.R. ... D b 12
Sandwich Harb., ...
Sandwich Harb., G.S.W.A. ... A f 3
Sandy Point, C.C. ... E g 10
Sandy Point, Ny. ... C c 16
Sangone Bay, P.E.A. ... E c 16
Sangura (Hatoka), P.E.A. ... C d 16
Sangura (Mavis R.), P.E.A. ... C d 16
Sandia, P.E.A. ... F a 15
Sanie, Bech. ... C c 4
Sama's Post, O.F.S. ... F b 9
Sanyara, Mash. ... F b 15
Sanyati, R., S.Z. ... D b 15
Saugani, B.C.A. ... E d 16
Sapa Sapa, C.F.S. ... B b 16
Saremoneng, C.C. ... D b 7
Sarmento, P.E.A. ... D c 16
Saron, C.C. ... D f 8
Sarua, P.E.A. ... B c 12
Sarna, R., Mash. ... B c 12
Sasin Koro, G.S.W.A. ... A c 15
Sasseb, G.S.W.A. ... A a 4
Sauls Kloof, S.A.R. ... B e 12
Sauls Kraal, S.A.R. ... E g 10
Saulsport, S.A.R. ... C d 13
Saurahelo, G.S.W.A. ... A b 4
Sawisis, G.S.W.A. ... A a 7
Schaapkuil, S.A.R. ... B c 12
Schildpadfontein, S.A.R. ... B c 12
Schildpadkop, C.C. ... C g 9

Schoeman's Drift, O.F.S.	F a	7	Shité Highlands,The,Ny.	D d	16	South Barrow, Nat	D e	10
Schoeman's Hoek, C.C...	B f	9	Shiré R., Ny.	D d	16	Southern Zambesia, S.A.	G a	4
Schoenberg, I.C.A.	B f	9	Shirozzi, P.E.A.	D d	16	Southeyville, C.C.	G d	9
Schoen Spruit, S.A.R...	F a	7	Shirwa, Kale, Ny	D d	16	South Sand Bluff, C.C...	D f	10
Schombie, C.C.	K d	9	Shishilaba, S.Z.	C h	15	Souvenir, Le, O.F.S.	G b	9
Schoon R., O.F.S.	D f	13	Shitambara, P.E.A.	C d	16	Spaldings, C.C.	B c	4
Schoons Spruit, S.A.R.	B e	13	Shitimba, Mash.	B b	4	Spandiikon, Nat.	D c	10
Schoorsteen, Berg, The,			Shitinclir Marsh, P.E.A.	C d	16	Spekakel, C.C.	B b	5
C.C.	C e	0	Shitunku, B.C.A.	C c	16	Spekheom R., S.A.R.	F c	13
Schulenburg, S.A.R.	B c	13	Shoa Lake, B.C.A.	B c	16	Spion Berg, The, C.C.	D c	8
Schulpfontein Point,			Shomali, Mash.	F b	15	Spion Berg, The, Nat.	G b	7
C.C.	B c	8	Shonni Saltpan, Bech...	C d	15	Spionkop, The, C.C.	D d	9
Schurede Mis., Bech.	B a	9	Shosha, P.E.A.	C d	16	Spitzkop Rand, S.A.R.	F b	13
Schweizerreineke,			Shoshona's, S.Z.	C b	15	Spitzkop, The, Bech...	F n	8
S.A.R.	A f	13	Shoshong, Bech.	C b	4	Spitzkop, The, C.C.	B d	9
Schwiezer, S.A.R.		13A	Shua R., Bech.	C d	16	Spitzkop, The, O.F.S.	E d	9
Scorpion Kraal, C.C.	C d	9	Shuitklip, C.C.	B b	7	Spitzkop, The, S.A.R.	D e	12
Scotland, The, C.C.	D e	8	Schulpfontein Point, C.C.	A e	7	Spoeg R., C.C.	B c	8
Scots Drift, Bech.	B e	4	Shumba, Mash.	K d	15	Springbock Vley, Bech.	B e	13
Scottsburg, Nat.	D o	10	Shungani, Mt., S.Z.	D b	15	Springbokfontein, C.C..	B b	8
Sea and Green Points			Shupanga, P.E.A.	D d	16	Springbok Kuil R., C.C.	C e	8
Light-House, C.C.	C f	8	Shuye, Bech.	B b	12	Springbok Vlakte, S.A.R.	D d	13
Seal, Cape, C.C.	D e	7	Sinbenze, B.C.A.	B d	16	Springfield, O.F.S.	B e	10
Seal I., C.C.	B e	7	Sibabariza, Mt., Mash.	E c	15	Springlontein, O.F.S.	E c	9
Seal Point, C.C.	C g	0	Sibai Lake, Tong.	F b	10	Springs, S.A.R.	D c	13
Seate R., Bas.	B d	10	Sibanani, Mat.	C c	15	Springvale, Nat.	D c	10
Sebolane, S.A.R.	D h	12	Sibatoul, Bech.	A e	12	Spruit Zonder Dr., O.F.S.	E f	13
Sebuhungi, Bech.	B h	4	Sibonia, Tong.	F b	10	Spuigkand, C.C.	B d	7
Sebungo, R., S.Z.	C c	15	Sicato R., S.A.R.	C d	13	Stanberton, S.A.R.	E c	13
Seebeli's Kingdom, Bech.	B b	4	Siefa, P.E.A.	F d	15	Stanford, C.C.	D g	8
Secuil, Bech.	B b	4	Sifumbat, P.E.A.	D b	4	Stanford, S.A.R.	C c	12
Secis, G.S.W.A.	A b	8	Sihambe, P.E.A.	E b	12	Stanger, Nat.	E d	10
Sefolul, S.A.R.	D c	12	Sikomana, P.E.A.	E c	12	Steelpoort, S.A.R.	D c	4
Sefulu, B.C.A.	C e	3	Sikwalakwala, S.A.R.	G a	13	Steenberg, S.A.R.	D d	13
Sehubia, Bech.	A d	12	Sikwane Hills, S.A.	B c	12	Steenkamps Berg,S.A.R.	E d	13
Sekate, Ny.	C c	16	Silooh, Bas.	F b	7	Steenkamp's Poort, C.C.	F e	8
Sekntewayo, S.A.R.	D d	12	Silube, Ny.	C c		Steenkool R., S.A.R.	E c	13
Sekhosi, B.C.A.	A d	16	Siluvu Hills, P.E.A.		3A	Steinkopf, C.C.	B b	8
Sekupalwa, Bech.	A c	15	Silva I., P.E.A.	D d	16	Steinkop R., C.C.	D b	7
Sekwati's, S.A.R.	E c	13	Silverton, S.A.R.	D d	13	Steinberg, ace Steyns-		
Selati Gold Field, S.A.R.	E d	13	Sinn, R., S.A.	A a	3	burg.		
Selati R., S.A.R.	F b	13	Simarianga, B.C.A.	C b	15	Steintaal, C.C.	B d	7
Selindeho, Bech.	B h	12	Sinzbo R., Mash.	E c	15	Stellaland, Bech.	A c	7
Selole, B.C.A.	F e	15	Simonstown, C.C.	C g	8	Stellenbosch, C.C.	C f	8
Selous's Road, P.E.A.	F e	15	Simsona Gold Field,			Stephanus Church, C.C.	F a	4
Sematali, Mat.	D d	15	Mash.	E b	15	Sterk R., S.A.R.	D e	13
Semalembuo, B.C.A.	B d	16	Sinane, B.C.A.	F b	13	Sterkstroom, Nat.	E c	10
Semeno R., Bas.	B d	10	Singwedsi R., S.A.R.	F b	13	Sterkstroom R. (Rusten-		
Semokwe, R., Mat.	C d	15	Singwezani R., S.A.R.	G b	13	burg), S.A.R.	C d	13
Semnkhu, P.E.A.	K c	16	Sinjoro, P.E.A.		3A	Sterkstroom R. (Waters-		
Sena, P.E.A.	C d	16	Sinkopie, Zul.	E c	12	berg), S.A.R.	C c	13
Senokal, C.C.	E c	9	Sinkolo, Tong.	E b	15	Sterkstroom R. (Zout-		
Sengoma, S.A.R.	B c	12	Sinein's, Mash.	E b	15	panaberg), S.A.R.	F a	13
Sengwe R., Bas.	D b	15	Sintilla, B.C.A.	C b	16	Stevenson Road,The, Ny.	C b	16
Senkunyane R., Bas.	B d	10	Sioux, B.C.A.	A d	16	Stewart, Mt., C.C.	E d	7
Senku, R., Bas.	C d	16	Sir Lowrie's Pass, C.C.	C g	8	Steyn, S.A.R.	D d	12
Sepatane R., S.A.R.	D b	13	Siserki, P.E.A.	F g	15	Steynsburg, C.C.	E d	9
Sephton, S.A.R.	B e	12	Sitanda, N.Z.	B c	10	Steynsdorp, SW.	G e	13
Seplan, C.C.	G d	9	Sitters Vley, C.C.	E h	8	Steytlerville, C.C.	D f	9
Sepu, R., P.W.A.	A d	16	Situbl, P.E.A.	F d	15	Stink Fontein, S.A.R.	B b	4
Sequati's Kraal, C.C.	G c	7	Sitwande's, P.E.A.			Stinkfontein, C.C.	C c	13
Serrbome, S.A.R.	D e	12	Six Mile R., S.A.R.	D d	13	Stockenstroom, dist., C.C.	F e	9
Serorome R., Bech.	B b	12	Sizgi Berg, The, C.C.	C d	9	Stolz, S.A.R.	D e	12
Serotti, Bech.	C e	15	Slangapies Berg, S.A.R.	F f	13	Stolzenfels, G.S.W.A.	B b	7
Serule R., Bech.	C e	15	Slangasa Range, Zul.	E c	16	Stompneus Bay, C.C.	B g	10
Seshcke, B.C.A.	B b	15	Slang Bay, C.C.	E e	7	Stormberg, The, C.C.	E d	9
Setlngoil, Bech.	A e	12	Slang Bergen, The, C.C.	E d	4	Stormberg Spruit, C.C.	F e	9
Setlngoil R., Bech.	A e	12	Slangkop Point, C.C.	C g	8	Storm R., C.C.	C e	9
Setuli, S.A.R.	B b	13	Slang R., C.C.	B f	10	Strand fontein Point, C.C.	B c	8
Sotomtsio, S.Z.	E d	15	Sledmere Flats, C.C.	G e	9	Strant Berg, The	F b	8
Seven Weeks' Pass, C.C.	F f	8	Slingerfontein, C.C.	D d	8	Streems Bay, C.C.	E g	8
Sewans, S.A.R.	D b	12	Slypsteln R., S.A R.	E a	13	Strays Bay, C.C.	E g	8
Seymour, C.C.	F e	9	Smit (Ilmstainburg),			Stresiger, Mt., C.C.	C f	9
Shahwe, P.E.A.	B c	16	S.A.R.			Strknkon, Bas.	E d	10
Shakha Badda, Mat.	C c	15	Smit (Zoutpansberg),			Strydom (Pretoria),		
Sha Nkomgo, B.C.A.	A c	16	S.A.R.	C b	12	S.A.R.	C c	12
Shamo, P.E.A.	D d	16	Smithfield, O.F.S.	E c	9	Strydom (Middelburg),		
Shamla, S.Z.	C c	15	Smithsburg, S.A.R.	E c	13	S.A.R.	D c	12
Shancng R., Bech.	C d	15	Snake R., G.S.W.A.	A c	4	Strydpoort Rand, S.A.R.	E c	13
Shanga, P.E.A.	E c	16	Sneeuw Bergen, The, C.C.	D d	9	Stryd R., C.C.	D c	9
Shangani R., Mat.	C c	15	Sneeuwkop, The, C.C.	B d	7	Stuartstown, Nat.	D e	10
Shangea, Mat.	C c	15	Snyders Fontein, C.C.	D b	7	Sturmans Pit, C.C.	B e	4
Sharpe, Fort, Ny.	C d	16	Snyman, Bech.	A c	16	Sudbury, C.C.	F g	9
Shashani R., Mat.	D d	15	Soana Ganga, C.F.S.	A b	16	Sugarloaf Rock	C f	9
Shashe R., Mash.	E c	15	Soana Mulopo, P.W.A.	A c	16	Suikerbosch Kop, Great,		
Shashi R., Mash.	D d	15	Sox Salt Pan, Bech.	C d	15	S.A.R.	F d	13
Shawbury, C.C.	H f	10	Soba Game, B.C.A.	B c	16	Sukene, S.A.R.	D b	12
Shephende, B.C.A.	E e	16	Sobuza Tr., Nat.	D d	10	Sullivan, Mt., C.C.	D c	9
Shepnerdson, S.A.R.	E a	7	Soco Reefs, C.C.	A n	8	Sumago, Bech.	A e	15
Shepstone, Port, Nat.	D o	10	Sofala, P.E.A.	E f	3	Sumba, P.E.A.	C e	16
Sherborne, C.C.	E d	9	Sohaap Vlei, C.C.	B d	8	Sumkel, Zul.	E c	12
Shesa R., B.C.A.	B d	16	Somerset East, C.C.	E e	9	Sumdump Tr.,G.S.W.A.	B a	4
Shesheka, B.C.A.	A d	16	Somerset West, C.C.	C g	8	Sunday R., C.C.	E e	9
Shidiani Tr., P.E.A.	E a	15	Somerset West Strand,			Sunday R., Nat.	D e	10
Shietmakar, O.F.S.	E b	7	C.C.	C g	8	Sunday River Pass, S.A.	C b	10
Shigenge, Ny.	C c	16	Somkodi Tr., Zul.	F c	16	Sunta, R., Bech.	A c	15
Shiketa, P.E.A.	F c	16	Somnos Water, C.C.	B c	8	Suu, S.Z.	E b	15
Shikumbuh, B.C.A.	A d	16	Songue R., Mat.	E c	16	Susoza, P.E.A.	C c	16
Shikunlo, P.E.A.	A c	16	Songwe, Ny.	C b	16	Sutherland, C.C.	D d	8
Shiloh, Bas.	A e	10	Soncab, G.S.W.A.	A a	7	Sutherland Hills, S.A.R.	D e	4
Shiloh, C.C.	F o	9	Sordwana, Port, Tong.	F b	10	Sutherland Hills, S.A.R.	E c	16
Shiloh, Mat.	D c	15	Sordwana Roads, Tong.	F b	10	Swalo, P.E.A.	E c	16
Shiluwane, S.A.R.	F c	13	Sorissa Point, P.E.A.	F c	16	Swakop R., G.S.W.A.	A a	7
Shinluwe, P.E.A.	D d	16	Soshe, P.E.A.	C d	16	Swampspoct, S.A.R.	D c	13
Shinto Kapenda, B.C.A.	C c	16	Sotai, C.F.S.	B c	16	Swart, S.A.R.	D d	12
Shipuriro, S.Z.	C b	15	Sources, The Mont aux,			Swartkand, Bas.	E d	10
Shinamamai, Mt.,P.E.A.	F c	15	S.A.	G b	7	Swaziland, S.A.	G c	13
Shinhwa, P.E.A.	C d	16	South African Republic,			Swellendam, C.C.	E g	8
Shinoya, see Chinolo.			S.A.	C c	12	Swellendam Point, C.C.	C g	8
						Tlotse, Bas.	B c	10

T

Taaiboschfontein, C.C.	C e	9
Taabachen, Mt., B.C.A.	B d	16
Table Bay, C.C.	C f	8
Table Mountain, C.C.	C f	8
Tabuka, Mat.	E c	15
Tacoma, P.E.A.	C d	16
Tafelberg Sta., C.C.	E d	9
Tafelberg,The Klein,C.C.	B e	13
Tafelkop, The, C.C.	D a	9
Tafelkop, The, O.F.S.	C h	10
Tafelskop (Lydenburg),		
The, S.A.R.	F d	13
Tafelskop (Potchef-		
stroom), The, S.A.R.	B e	13
Taiasket, G.S.W.A.	A a	4
Taibosch R., O.F.S.	C e	13
Takun, Bech.	B c	4
Takwaning, Bech.	E a	7
Tamalukan R., Bech.	A e	15
Tama Malisa, Mat.	C c	15
Tambala, Ny.	C c	16
Tambo Akilala, B.C.A.	B c	16
Tambooti R., S.A.R.	C b	13
Tambasi I., P.E.A.	E c	16
Tandtjesberge, The, C.C.	E e	9
Tanqua R., C.C.	E e	8
Tapomanda,Cape,P.E.A.	C c	9
Tapila, C.C.	C c	9
Tarka R., C.C.	F c	9
Tarkastad, C.C.	F d	9
Tatam Magha, Bech.	B d	15
Taras Berg, The, C.C.	B a	8
Tati, dist., Bech.	C d	15
Tati, P.E.A.		3A
Tati R., Bech.	C d	15
Tati R., Bech.	A e	12
Taungs, Bech.	B c	4
Tauopa, Bech.	C c	15
Tave R., S.A.R.	D b	12
Tekomaji I., P.E.A.	E c	16
Telingana, S.A.	D b	12
Telle R., S.A.	A e	10
Tembe, R., P.E.A.	E d	12
Tembuland, C.C.	A f	10
Tembwe, Ny.	C c	16
Temby R., P.E.A.	E d	16
Tenedos, Port, Zul.	E d	10
Tenedos Shoal, Zul.	E d	10
Tenke, N., B.C.A.	B c	16
Terblans, S.A.R.	G a	7
Teresa, Ny.	C c	16
Terne R., P.E.A.	E c	12
Tete, P.E.A.	C d	16
Teyateyaneng, Bas.	A d	10
Thaba Bosigo, Bas.	A d	10
Thaba Enzinhe Hill,		
Mash.	E c	15
Thabana Morena, Bas.	A d	10
Thabancho, O.F.S.	F b	9
Thaba Patchoa, O.F.S.	F b	9
Thaba Ncio, S.A.R.	F e	13
Thabine R., S.A.R.	F c	13
Theo R., S.A.R.	F e	13
Thelo, Bas.	A d	10
Thelesn R., S.A.R.	F c	13
Theopolis, C.C.	E f	9
Thesiger, Mt., C.C.	C f	9
Thikhando, Bas.	B d	10
Thlotsi, Bas.	C c	4
Thokwe R., S.A.R.	C e	12
Thomas, Bas.	G e	7
Thomas Dreyer Berg,		
S.A.R.	B c	13
Thompson, C.C.	E a	7
Thorn Bay Point, C.C.	C d	8
Thorndale, S.A.R.	B c	12
Thorngrove, C.C.	E e	9
Thornhill, C.C.	E b	7
Thousand Pools, Land of		
the, Mat.	C e	15
Three Sisters Sta., C.C.	C d	9
Three Sisters, The,S.A.R.	B c	13
Thys Bay, C.C.	E c	7
Tibil, P.E.A.	F d	15
Tiger Berg (Aberdeen),		
The, C.C.	D e	9
Tiger Berg (Calvinia),		
The, C.C.	E d	8
Tiger Berg (Namaqua-		
land), The, C.C.	B c	8
Tiger Kloof, The, C.C.	E c	7
Tiger Kloof Spruit,O.F.S.	B e	13
Timbaland R., S.A.R.	G c	13
Timbane, P.E.A.	F d	15
Tina, C.C.	B e	10
Tina R., C.C.	B e	10
Tinge R., Bech.	B c	16
Tingua, B.C.A.	C b	16
Tlakamamo, Bech.	B e	4
Tlambeli, Bech.	C d	15
Tlaping Spruit, Bech.	A e	12
Tlotse, Bas.	B c	10

Tlotse R., Bas. Bd 10
T'Nous, C.C. Bb 7
Toa R., P.E.A. 3A
Tobos, G.S.W.A. Ba 7
Toestaan, C.C. Be 9
Tofikey, Ny. Cb 16
Tokanna R., Bech. Be 4
Tokoji, Bech. Be 4
Tokwe, R., Mash. Ed 15
Tola, P.E.A. De 16
Toleni, C.C. Bg 10
Tolo Azime Falls (Limpopo R.), S.A. Ee 15
Tonders, C.C. Bd 7
Tongaati R., Nat. Ed 10
Tongaland, S.A. Fa 10
Tonk, Bech. Da 7
Tonko R., Bech. Ac 15
Toorn Berg, The Groote, C.C. Dd 8
Toorns R., Klein, C.C. Dd 8
Toro, C.C. Bf 10
Totela, C.F.S. Bb 16
Touns Berg, C.C. Ba 8
Touw R., C.C. Ef 8
Touws, The, C.C. Ef 8
Touws River Sta., C.C. Ef 8
Tradouws Pass, C.C. Ef 8
Traka R. (Prince Albert), C.C. Bf 9
Traka R. (Worcester), C.C. Ef 8
Traka, The, C.C. Cf 9
Transkei, The, C.C. Bg 10
Transvaal, The, S.A. Be 12
Trekveld, The, C.C. De 8
Triangle, C.C. Df 8
Troe-Troe, C.C. Ad 4
Tromskop, The, C.C. Co 7
Troyeville, S.A.R. 13A
Tsakoma, S.A.R. Fb 13
Tsaun, G.S.W.A. Ab 4
Tsande R., S.A.R. Gh 13
Tsening, Bech. Da 7
Tshungwana, C.C. Bo 10
Tshwani, S.A.R. Be 12
Tsitano Saltpan, Bech. Cd 15
Tsitsa Falls (Tsitsa R.), C.C. Bf 10
Tsitsa R., C.C. Bf 10
Tsojana, C.C. Fo 7
Tsolo, C.C. Ge 7
Tsolo, Old, C.C. Bf 10
Tsomo, C.C. Ag 10
Tsomo R., C.C. Af 10
Tsumis, G.S.W.A. Ab 4
Tswaing R., Bas. Ad 10
Tugela R., S.A. Dc 10
Tugela R., Little, Nat. Ce 10
Tugulu, P.E.A. Ec 16
Tuin, C.C. Cd 7
Tulbagh, C.C. Df 8
Tuli, Mat. Dd 15
Tuli, R., Mat. Dd 15
Tumbe, P.E.A. De 16
Tundalanga, C.F.S. Ab 16
Tungi, P.E.A. Ec 16
Tutluane, S.A.R. Cb 12
Tutuan, Tong. Ed 12
Tungwisi R., Mash. Fe 15
Tuuxa R., C.C. Ge 9
Twass, G.S.W.A. Ab 4
Tweeddale, C.C. Ed 9
Tweede Bergen, The, S.A.R. Ea 13
Tweede Poort, S.A.R. Be 12
Twee Mik Berg, The, C.C. Fe 8
Twenty-four Rivers, C.C. Df 8
Twins, The, C.C. Be 8
Twist Niet, C.C. Fd 9
Tylden, C.C. Ge 9
Tylden Peak, C.C. Fd 9
Tyotyo, R., B.C.A. Cb 15

U

UANGU, P.E.A. De 16
Ubazi R., C.C. Cf 10
Ubip R., C.C. Aa 8
Ubomboo Head, C.C. Cf 10
Uchungu, dist., E.A. Cb 16
Ugie, C.C. Ge 7
Ugrabib Berg, The, C.C. Cb 9
Ugrabis, C.C. Bb 8
Ugweno, co., Nat. Dd 10
Uhabis, G.S.W.A. Ab 7
Uilkraals Bay, C.C. Dg 8
Uilkraals R., C.C. Dg 8
Uisip, Bech. Cb 7
Uitdraai, O.F.S. Eb 9
Uitenhage, C.C. Ef 9
Uitkyk, C.C. Ea 7
Uitspan Berg, The, C.C. Ce 8
Ukula, Mt., P.E.A. Dc 16
Ukwampa R., Mat. De 15
U'Larkeni Drift, P.E.A. Fd 15
Ulenji, B.C.A. Da 15
Ulundi, Nat. Cd 10
Ulundi, Zul. Ec 10
Umab Desert, G.S.W.A. Bb 4
Umalooy, Sw. Dd 12
Umbelosi R., E.A. Ed 12
Umbelosi R., Black, Sw. Ge 13
Umbelosi R., White, Sw. Ge 13
Umbigiza, Tong. Fa 10
Umchabanchaban, P.E.A. Ed 12
Umchanatsi, S.Z. Ed 15
Umchengwisi R., P.E.A. Eb 12
Umchingwe R., Mat. Dd 15
Umfanawenlela, Zul. Do 12
Umfuli Gold Field, Lower, Mash. Db 15
Umfuli Gold Field, Upper, Mash. Ec 15
Umfuli, R., Mash. Ee 15
Umfunge, R., S.Z. Eb 15
Umgalunguio Mt., Mat. Ee 15
Umgazi R., C.C. Ge 7
Umgazwi, P.E.A. Fd 15
Umgeni, Nat. Ed 10
Umgesi R., Mash. Ec 15
Umgitywa, Zul. De 12
Umgorim R., P.E.A. Fe 15
Umgova Mts., Zul. Ee 10
Umgovuma R., Sw. Eh 10
Umguasi, R., Mash. Eb 15
Umgwangwana R., C.C. Gb 7
Umgwenia, P.E.A. Db 4
Umhlanganini, R., P.E.A. Fd 15
Umhlangen, Mat. Dc 15
Umhlatoos R., Sw. Ea 10
Umhloti R., Nat. Dd 10
Umjindi, Tong. Fb 10
Umkhosi, R., Mat. Ce 15
Umkobowan, Sw. Dd 12
Umkof, P.E.A. Ed 12
Ukmomaas R., Nat. De 10
Umkomaas R., Sw. Ge 13
Umkomanzi, co., Nat. Dd 10
Umkomanzi R., Nat. De 10
Umkombies R., S.A. Fo 13
Umkonto R., Sw. Dd 12
Umkoshloli R., E.A. Db 4
Umkoshloli R., P.E.A. Eb 12
Umkumbura, R., P.E.A. Ea 15
Umkusi R., S.E.A. Eb 10
Umlandela, Zul. De 12
Umlandela Tr., Zul. Ec 10
Umlatusi R., Zul. Ec 10
Umlazi R., Nat. Dc 10
Umnyati, Mash. Ec 15
Umnyati R., Mash. Db 15
Umona R., Zul. Ec 10
Umpambinyoni R., Nat. De 10
Umpumulu, Nat. Ed 10
Umqnakela, C.C. Cf 10
Umsaugaadzi R., P.E.A. 8A
Umsawo R., Mat. Ed 15
Umsengaisi R., E.A. Eb 15
Umshabetsi R., Mat. Dd 15
Umshagashi R., Mat. Db 4
Umsikaba R., C.C. Cf 10
Umsinga, Nat. Dc 10
Umsinga, Mt., Nat. Dc 10
Umslane R., Mat. Dd 15
Umsunze, Bech. Cd 15
Umswaas, P.E.A. Ed 12
Umtagat, Sw. Dd 12
Umtali, Mash. Fc 15
Umtamvuna R., S.A. Df 10
Umtanbeka, Sw. Eb 10
Umtanga, Mt., Mat. Ec 15
Umtasa's, P.E.A. Fe 15
Umtasiti R., S.A.R. Dc 12
Umtata, C.C. Ge 7
Umtata, C.C. Bf 10
Umtata R., C.C. Cf 10
Umtegan, Mat. Dd 15
Umtentu, C.C. Bf 10
Umtentu R., C.C. Df 10
Umtigesa, Mash. Ec 15
Umtlitlo R., S.A.R. Ge 13
Umtshefu R., P.E.A. Fe 15
Umtule R., S.A.R. Fe 13
Umtwalumi R., Nat. De 10
Umvolosi R., Black, Zul. Ee 10
Umvolosi R., White, S.A.R. Ee 10
Umvoti, co., Nat. Dd 10
Umvoti R., Nat. Dd 10
Umvukwe Mts., Mash. Eb 15
Umvule, Zul. Ee 10
Umvulo R., Zul. De 12
Umyangu, P.E.A. Dc 4
Umzimhlavana R., C.C. Ge 7
Umzimkulu, C.C. Ce 10
Umzimkulu R., Nat. Cd 10
Umzimlava R., C.C. Ge 7
Umzimvubu R., C.C. Ce 10
Umzingwane, Mat. Dd 15
Umzingwane R., Mat. Dd 15
Umzinto, Nat. De 10
Umzinto R., S.A.R. Gd 13
Umzinyati Lake, Zul. Fe 10
Umzinyati R., S.A. Db 10
Umzumbi, Nat. De 10
Umzumbo R., Nat. De 10
Umzwaas, S.A.R. Db 12
Umzweswie Gold Field, Mash. Dc 15
Umzweswie, R., Mash. Dc 15
Undakalyi R., C.C. Cf 10
Unde, P.E.A. Dc 16
Undi, P.E.A. Ce 16
Undungaswe, P.E.A. Db 4
Unguesi, R. (R. Kafue), B.C.A. Bd 16
Unguesi R. (R. Zambesi), B.C.A. Bd 16
Ungwali, C.C. Ag 10
Ungwenia, P.E.A. Ec 12
Uniondale, C.C. Cf 9
Union Vley, Bech. Bh 4
Unodwengo, Zul. Du 12
Unyameni R., C.C. Df 10
Unyango, P.E.A. Dc 16
Unyanyene R., S.A.R. Dc 10
Upamba Lake, C.F.S. Bb 16
Upa, R., E.A. Fb 15
Upindo Tr., C.C. Be 10
Upington, Bech. Cg 3
Upper Tugela, town, Nat. Ce 10
Upper Umfuli Gold Field, Mash. Ec 15
Upper Zak R., C.C. Ed 8
Urema R., P.E.A. 8A
Uridanab, G.S.W.A. Ba 7
Urigab, G.S.W.A. Ab 4
Urinoub R., C.C. Ea 8
Urua, dist., C.F.S. Bb 16
Urungu, dist., B.C.A. Cb 16
Us, G.S.W.A. Aa 7
Useme, Zul. Dc 4
Useme, Zul. De 12
Usibebu's Kraal, Zul. Eb 10
Usimelo, P.E.A. Ed 12
Usonia, P.E.A. Ce 16
Ussambi Tr., C.F.S. Bb 16
Usuto Tr., Zul. Eb 10
Usutu R., P.E.A. Ed 12
Usutu R., Great, Sw. Ge 13
Usutu R., Little, Sw. Ge 13
Utabi, R., S.A.R. Db 12
Utale, Ny. Cd 16
Utrecht, S.A.R. Db 10
Utshaniud, P.E.A. Fd 15
Uvula's, Mat. Dd 15
Uwiwa, dist., Ny. Cb 16
Uxamaris, G.S.W.A. Ba 7

V

VAALHEUVEL, GROOT, C.C. Cb 8
Vaalheuvel, Klein, C.C. Cb 8
Vaalkop, The, S.A.R. Be 13
Vaal R., S.A. Ga 7
Vaal R. (Namaqualand), C.C. Ab 7
Vaalwater R., S.A.R. Fe 13
Vacca, Cape, C.C. De 7
Vai R., C.C. Cd 8
Valdezia, S.A.R. Fb 13
Valsh R., O.F.S. Ab 10
Van Reenen's Pass, S.A. Ce 10
Van Rhynsdorp, C.C. Cd 8
Van Wijk's Vlei, C.C. Fe 8
Van Wyk, S.A.R. Bd 12
Vechtkop, The, O.F.S. Bb 10
Venten R., O.F.S. Ef 13
Venter, S.A.R. Ce 12
Venters, S.A.R. Db 12
Ventersburg, C.C. Fc 7
Ventersburg, O.F.S. Fb 7
Ventersdorp, S.A.R. Be 13
Venterskroon, S.A.R. Ce 13
Venterstad, C.C. Ee 9
Vereeniging, S.A.R. Ce 13
Vermaak, S.A.R. Dd 13
Vermaak, Sw. Eb 10
Verloren R., C.C. Ce 8
Verulam, Nat. Ed 10
Verzamel Berg, The, S.A.R. Ff 13
Vetberg, C.C. Db 8
Vet R., O.F.S. Fa 7
Vette R., C.C. Fg 8
Vicenti, P.E.A. Fe 3
Victoria, Mat. Ed 15
Victoria, Nat. Ed 10
Victoria East, dist., C.C. Fe 9
Victoria Falls (R. Zambesi) Bb 15
Victoria, Fort, S.A.R. Ee 10
Victoria Gold Field, Mat Ed 15
Victoria Mine, C.C. Bb 7
Victoria West, C.C. Bd 9
Victoria West Sta., C.C. Cd 9
Vidal, Cape, Tong. Fe 10
Viljoen, S.A.R. Dc 12
Viljoen's Drift, S.A.R. Ce 13
Villiersdorp, C.C. Dg 8
Villiersdorp, O.F.S. Df 13
Vischwater R., S.A.R. Eb 13
Vlei R., C.C. Bd 4
Vlugt Kraal, S.A.R. Fd 13
Vogel Fontein, C.C. Cd 7
Vogel Klip, The, C.C. Bb 8
Vogel R., C.C. Ee 9
Vogel Vallei Vloer, C.C. Ec 8
Volker, S.A.R. Eh 10
Volksrust, S.A.R. Ef 13
Voltas, Cape, C.C. Ae 4
Vondeling I., C.C. Ad 7
Vorster, S.A.R. Ce 12
Vrede, O.F.S. Cb 10
Vredefoot, O.F.S. Fa 7
Vredendal, C.C. Be 7
Vrijheid, S.A.R. Db 10
Vryburg, Bech. Be 4
Vryheid, *see* Vrijheid.
Vunga, P.E.A. Fb 15
Vurmelo, Mat. Ed 15
Vurruca, P.E.A. Dd 10
Vuurdood, The, C.C. Ga 8

W

WAAI FONTEIN, C.C. De 7
Waal Hoek, C.C. Fd 8
Wagenaars Kraal, C.C. Bd 9
Wahlberg, Bech. Bb 4
Wahode, P.E.A. Hh 13
Wakkerstroom, S.A.R. Ff 13
Walfisch Bay, *see* Walvisch Bay.
Walker Bay, C.C. Dg 8
Walker I., P.E.A. Ed 10
Waller, Mt., Ny. Ce 16
Walker Point, C.C. Bg 9
Wallmansthal, S.A.R. Dd 13
Walthoorns Kraal, C.C. Ce 9
Walvisch Bay, S.W.A. Af 9
Wamisi I., P.E.A. Ec 16
Wandonde Tr., P.E.A. Dc 16
Wankie, B.C.A. Cb 15
Wanetze, Mat. Ed 15
Warden, Fort, C.C. Bg 10
Warmbad, G.S.W.A. Bb 7
Warmbath, S.A.R. Fd 13
Warm Bokkeveld, C.C. Df 8
Warmwater Bergen, C.C. Ef 8
Warrenton, C.C. Da 9
Warwick, Fort, S.A.R. Dc 10
Waschbank, Nat. Dc 10
Waschbank, Peak, C.C. Gd 9
Water Berg, The, S.A.R. Ce 12
Waterfall Bluff, C.C. Cf 10
Waterfall R., S.A.R. Ga 7
Watergras Drift, C.C. Ab 7
Waterloo Bay, C.C. Gf 9
Watersberg, dist., S.A.R. Ch 13
Waterval R., S.A.R. Dc 13
Weber, Fort, S.A.R. Dc 12
Wedge Point, C.C. Ab 7
Wedza Mt., Mat. Ec 15
Weeber, Fort, S.A.R. Ec 13
Weenen, Nat. Ce 10
Wegloop, Tati Cd 15
Welcome Berg, The, C.C. Be 7
Wellington, C.C. Df 8
Weltevreden, C.C. Ge 7
Wemmer Vlei, S.A.R. 13A
Wepener, O.F.S. Fb 9
Wesley, G.S.W.A. Ab 4
Wessels Nek Sta., Nat. Ce 10
Weston, Nat. Dd 10
Wetterhorn, The, C.C. Ae 10
Whale Rock, C.C. Cf 10
Wheeler, S.A.R. Dc 12
White Kei R., C.C. Fd 9
White Point, C.C. Ab 8
White Umvolosi R., S.A.R. Ee 10
Whittlesea, C.C. Fe 9
Wildebeest Pan, C.C. Cb 9
Wilge R., O.F.S. Ga 7
Wilge R., S.A.R. Dd 13
Wilhelms R., C.C. Dd 8
Wilkerhout's Drift, Bech. Cb 7
William, Fort, C.C. Ce 10
William, Fort, S.A.R. Ce 12
Williamstown, Nat. Ed 10
Willowmore, C.C. Cf 9
Willowvale, C.C. Bg 10
Winburg, O.F.S. Fb 7
Windhoek, C.C. Bb 8
Windhoek, G.S.W.A. Bf 3
Windvogel Mts., C.C. Fe 9
Winkledrift, O.F.S. Cf 13
Winter Berg, The, C.C. Be 9
Winterberg, The Great, C.C. Fe 9
Winter Hoek Mts., C.C. Ef 9
Winterhoek Mts., Klein, C.C. Df 9

Winter Hoek, The Great, C.C. ... Cf 8
Wintervold, The, C.C.... Cd 9
Witberg, C.C. .... Db 7
Witfontein Berge, S.A R Bc 13
Witklip, S.A.R. .... Eb 13
Witmoss, C.C. .... Ee 9
Witputs, C.C. .... Db 9
Witsanda, C.C.......... Ba 9
Wittebank, C.C......... Ab 7
Witte Bergen, The, O.F.S. ...... Bc 10
Wittebergen (Barkly), The, C.C. ... Ae 10
Witte Bergen (Griqualand West), The, C.C.. Be 10
Witte Elsbosch, town, C.C. .... Dg 9
Witte Elsbosch, The, C.C. .... Df 9
Witte Klip, The, C.C.. Bo 8
Wittewaters, C.C. ..... Co 8
Witvley, G.S.W.A. ..... Ab 4
Witwater, The, C.C. ... Ba 9
Witwatersrand, S.A.R... Ce 13
Wiekpoort R., C.C. .... Ed 9
Wodehouse, dist., C.C... Fd 9
Wolf Spruit, S.A.R. ... Bf 13
Wolf Poort R., C.C. ... Bc 8
Wolmarans, S.A.R. .... Bd 12
Wolmaranstad, S.A.R... Bf 13
Wolvefontein, C.C. .... Df 9
Wolve Spruit, O.F.S. .. Ea 9
Wonderbooms R., C.C... Fd 9
Wonderfontein, S.A.R... Bd 12
Wonderfontein Loop R., S.A.R. .... Ce 13
Wonderhauvel, C.C. ... Ec 7
Woodbush Gold Field, S.A.R. .... Eb 13
Woodside, O.F.S. ...... Bb 10
Woodville, C.C. .... Bf 9
Woody Cape, C.C. ..... Fg 9
Woolridge, C.C. .... Gf 9
Worcester, C.C. .... Df 8
Wreck Point, C.C. ..... Ab 7
Wupperthal, C.C........ Dv 8
Wynberg, C.C........... Cg 8

X
Xalanga, C.C. .... Af 10
Xamates, G.S.W.A. .... Aa 7
Xanob R., G.S.W.A. ... Aa 7
Xesibe Tr., C.C. ...... Ce 10
Xnabara R., C.C. ..... Gd 7
Xosa, C.C. .... Db 7
Xoungs, C.C........... Db 7
Xuka, C.C. .... Fc 7
Xuka R., C.C. .... Bf 10
Xurutabi's, G.S.W.A. .. Aa 7
Xutsa, C.C. .... Fd 7

Y
Yamkombe, R., Mash. .. Fb 15
Yango, P.E.A. .... Dd 16
Yankwest, Mash........ Fb 15
Yao Tr., P.E.A. .... De 16
Yeoville, S.A.R. 13a
Yolland, Fort, Zul...... Ec 10
York, Nat. .... Dd 10
Yzerberg, The, S.A.R. .. Fb 13

Z
Zak R., C.C. .... Bd 9
Zak R., Upper, C.C. ... Eb 8
Zambesia, Northern, S.A. Bo 16
Zambesia, Southern, S.A. Ca 4
Zambesi Delta, P.E.A. .. Dd 16
Zambesi, R., E.A. ..... Bd 16
Zambili, Tong. .... Ed 12
Zambot, Sw. .... Eb 10
Zand R., C.C. .... Cb 9
Zand R., O.F.S. ...... Ac 10
Zand R. (Watersberg), S.A.R. .... De 13
Zand R. (Zoutpansberg), S.A.R. .... Ea 13
Zand River Bergen, S.A.R. .... De 13
Zandveld, The, C.C. ... Cd 8

Zanre, dist., P.E.A. .... Fc 15
Zapaira, B.C.A. .... Cc 16
Zastron, O.F.S. .... Gc 9
Zcbano, S.A.R. .... Db 12
Zebedela's Kraal, S.A.R. Ec 13
Zeekoe Point, C.C. .... Ce 7
Zeerust, S.A.R. .... Ad 13
Zekoo R., C.C. .... De 9
Zeven Fontein Pan, C.C. De 8
Zibi, C.C. .... Bo 10
Zielsman, Nat.......... Ce 10
Zimbabwe, Mat......... Ed 15
Zimulu's, Mat......... Ba 4
Zingabila, Mash....... Eb 15
Zion, Bas. .... Ad 10
Zitzikamma Forest, C.C. Cf 9
Zitzikamma Point, C.C. Dg 9
Zin-zin R., P.E.A. .... Dd 16
Zoani, Ny............. Ce 16
Zoar, C.C. .... Ff 8
Zoasamoio, P.E.A...... Fc 15
Zoekoegat, C.C. ...... De 9
Zomba, Ny. .... Dd 16
Zombe, B.C.A......... Cb 16
Zonder Dr., Spruit,O.F.S. Ef 13
Zonder Einde Mts., C.C. Dg 8
Zonder Einde R., C.C. . Dg 8
Zongoro, Mash. ...... Fb 15
Zongwo, R., B.C.A. ... Cb 15
Zoning, C.C........... Ee 8
Zouga R., see Zuga R.
Zuur Berg, The, C.C.... Ed 9
Zout Pan, C.C......... Ch 9
Zoutpans Bergen, S.A.R. Eb 13
Zoutpansberg, S.A.R. .. Eb 13
Zoutpansberg, The, C.C. Ed 9
Zoutpans Drift, O.F.S. . Eb 7
Zout R. (Swellendam), C.C. .... Eg 8
Zout R. (Vanrhynsdorp), C.C. .... Hd 8
Zout River Vlei, C.C... Co 9
Zuga R., Bech......... Cf 8
Zuikerbosch Rand, S.A.R. .... De 13
Zuikerbosch, R., S.A.R. Ga 7
Zululand, S.E.A........ Ec 10

Zurobo, P.E.A......... Cd 16
Zundo, S.A.R. .... Ga 13
Zuurberg (Alexandria), The, C.C. ...... Ef 9
Zuur Berg (Griqualand East), The, C.C.... Ce 10
Zuurbeank, C.C. ...... Ef 8
Zuurpoort, C.C. ...... Dd 9
Zwaart Doorn R., C.C.. Bc 7
Zwaart Kop, C.C. ..... Ce 7
Zwagers Hoek, C.C. ... Eo 9
Zwakers Hoek, S.A.R. . Dc 13
Zwariberg, The, C.C.... Ce 10
Zwartbank, C.C. ...... Ab 7
Zwart Berg (Carnarvon), The, C.C. ...... Fb 8
Zwart Berg (Malmesbury), The, C.C. ... Cf 8
Zwart Berg, The Great, C.C. .... Cf 8
Zwart Doorn R., C.C... Cc 8
Zwart Bergen (Ceres), The, C.C. ...... Df 8
Zwarte Bergen (Prince Albert), The, C.C. ... Ff 8
Zwarte Bergen, The Great, C.C. ...... Af 9
Zwarte Bergen, The Little, C.C. ...... Bf 9
Zwarteberg Pass, C.C. . Ff 8
Zwarte Ruggens, The, C.C. .... Df 9
Zwart Kopples, O.F.S. . Fc 7
Zwart Koppies, S.A.R. . Cd 13
Zwartkops Junction, C.C. Ef 9
Zwartkops R., C.C. .... Ef 9
Zwartkop, The, C.C. ... Bb 8
Zwartland, C.C. ...... Cf 8
Zwart Lintjes R., C.C. . Bc 8
Zwart Modder, Bech.... Bc 4
Zwart Modder, C.C. ... Bb 7
Zwart R., C.C. .... De 9
Zwart R., C.C. .... Bg 9
Zwart Ruggens, The, S.A.R. .... Bd 13
Zwellenham, see Swellendam.
Zwingel Pan, C.C. ..... Cb 9

9 783337 117498